I0198929

NUECES BATTLE AND MASSACRE:
Myths and Facts

By

Wm. Paul Burrier, Sr.

Watercress Press
San Antonio, 2015

Wm. Paul Burrier

Copyrighted 2015

ALL RIGHTS RESERVED. No part of this book may be reproduced in any form without the written permission for the author, except for brief passages included in brief passages included in research papers, books, newspapers or magazines

A *Watercress Press* book
from Geron & Associates
www. watercresspress.com

ISBN-13: 978-0-9897822-1-0
Library of Congress Control Number: 2014959050

Cover image by Wm. Paul Burrier, Sr.
Cover design by 3iii's Graphic Studios

NUECES BATTLE AND MASSACRE:
Myths and Facts

Foreword
From the Author

The story of the Nueces Battle and Massacre which took place on August 10, 1862, at the West Prong of the Nueces River in Texas is full of mystery and myths. Over the years certain myths have arisen and been accepted as facts, with the result that nearly everything written about this event is either strongly biased, misinterpreted or flat-out incorrect.

Because of the magnitude of errors contained in most accounts of the event, I set out to write a corrected version. I realized that the reader might be overwhelmed by the data and references, so I decided to publish a series of what I called "little books," setting out my various sources and contemporary accounts. This is the fourth and last of these "little books." The previous three are August Siemering's *Die Deutchen in Texas Waehrend Des Buergerkrieges, (The Germans in Texas During the Civil War*, published by Llumina Press, 2013; *Confederate Military Commission Held in San Antonio, Texas, July 2—October 10, 1862*, and *Nueces Battle and Massacre Source Documents*, both published by Watercress Press, 2014.

Wm. Paul Burrier

Leakey, Texas
January, 2015

Table of Contents

1 – Opposition to Slavery

MYTH: The German settlers of the Texas Hill Country were totally opposed to the 'peculiar Southern institution', i.e., slavery. [1]

FACT: The German settlers, like most groups of Hill Country settlers, were neither totally opposed nor totally in favor of slavery. One of the more radical German settlers owned a slave. [2]

DISCUSSION: This is a continuing debate in the German-Texas intellectual community. On one side of the issue is Terry Jordan who says, "To call them (Hill Country Germans) abolitionists would be a serious misinterpretation of the facts." Jordan points out, "The German peasant of the nineteenth century was not a politically oriented being, and nothing could distort the picture more than to depict the average Texas German farmer in the western settlements as an active abolitionist." Even the abolitionist Frederick Olmsted in his visit to Texas in 1854 wrote, "Few of them (German settlers) concern themselves with the theoretical right or wrong of the institution, and while it does not interfere with their own liberty of progress, are careless of its existence." Terry is supported by Rudolph Biesele who says, "They (the German settlers) regarded slavery as an institution with which the federal government had nothing to do." To support this view Biesele points out that, "The largest slave-owner among the Germans in Texas was the Society for the Protection of German Immigrants in Texas, which had twenty-five slaves on the Nassau Farm in 1848 . . ."[3]

On the other side of the debate is Walter Kamphoefner, Professor of History at Texas A & M, who says that Jordan "goes too far in characterizing Texas Germans as unremarkable in their race attitudes, and . . . underestimates the degree to which Germans stood apart from their fellow Texas on the issue of the Civil War," but seems to agree most German settlers were not abolitionists. [4]

The problem was the Hill Country German population contained an ultra-radical group of Germans, Freethinkers (opposed to an organized church) and Forty-Eighters (who took part in the failed German Revolution of 1848-49). Both groups were extremely politically-minded and opposed to slavery. These Freethinkers and Forty-Eighters attempted to unite the Texas German communities under their leadership in 1854 at the San Antonio Convention. [5] They greatly influenced the anti-secession vote in the Texas Hill Country, especially in Gillespie County, and were the leaders of the Union Loyal League.

ENDNOTES - Myth #1

1. Biggers, Don H.; *German Pioneers In Texas*, (Fredericksburg Standard, Fredericksburg, Texas 1925), p. 57.

2. Jordan, Terry C., *German Seed in Texas Soil; Immigrant Farmer in Nineteenth-Century Texas*, (University of Texas Press, Austin, Texas 1966), pp. 109 & 182; Olmstead, Frederick Law, *A Journey Through Texas Or, A Saddle-trip on the Southwestern Frontier*, (University of Texas Press, 1978), p. 432; Biesele, Rudolph Leopold, *The History of the German Settlements in Texas 1831-186*, (Eakin Press, Austin, Texas, 1986), p. 196.

3. Kamphoefner, Walter D., 'New Perspectives on Texas Germans and the Confederacy', *Southwestern Historical Quarterly*, Volume CII No. 4 (April 1999), p. 179.

4. Biesele Rudolph L., 'The Texas State Convention of Germans in 1854', *Southwestern Historical Quarterly*, Volume XXXIII No 4, April 1930.

2 – The Texas Vote on Secession

MYTH: The Ordinance of Secession passed in Texas because of light voter turnout and by a small margin. Many pro-Union voters did not vote because they feared retaliation from the secessionists. [1]

FACT: The Ordinance of Secession passed by a majority of over 75%. Voter turnout was only 3-4% less than in the presidential election of November 1860. [2]

DISCUSSION: The total vote on the Ordinance of Secession in February 1861 was 61,337. A total of 46,188 voted for secession and 15,149 voted against. The total vote in the presidential election of November 1860 was 63,773. Thus only 2,436 fewer voters turned out for the 1861 Ordinance of Secession than for the 1860 presidential election. The Ordinance of Secession failed in the six counties of the Texas Hill Country (defined as Bandera, Blanco, Kerr, Gillespie, Medina, and those parts of Blanco, Comal and Kerr which now compose Kendall County) by a count of 361 for secession and 864 against. [3]

Studies show in the 1860 presidential election somewhere about 39% of the eligible voters did not vote. In the secession referendum of February, 1861, somewhere about 46% of the eligible voters did not vote. This was about 7% less voted in the February secession referendum than in the 1860 presidential election. The actual difference in voters is 2,436. If all 2,436 had voted against secession the results still would have over 72% for secession. Professor Dale Baum, of Texas A & M University, examined the question, "To what extent did intimidation of Texas Unionists and deliberate miscounting of votes affect the outcome of the February balloting?" He examined two previous statewide elections and developed a predicted secessionist percentage in each Texas county. His results established that up to 18% higher would be normal, based on several variables.

The results showed that only twenty-two of the state's one hundred twenty-two counties had a higher percent than the expected 18%. Grimes County was the single Texas county that had a gain of 'for' secession higher than 42%.

Professor Baum's study further showed that twenty-one Counties had a higher percent than the expected 18% 'against' secession, the highest being Uvalde with 69% followed by Montague and Gillespie with 53% Blanco with 51% and Medina with 39%. In summary the highest voting irregularities were in favor of the Unionists – not the secessionists. [4]

ENDNOTES - Myth #2

1. Sansom, John W., *Battle Of Nueces River In Kenney County, Texas August 10, 1862*, (privately published, San Antonio, Texas, 1905), p. 2; Biggers, Don H., *German Pioneers*, p. 57; Ransleben, Guido E., *A Hundred Years of Comfort in Texas; A Centennial History*, (The Naylor Company, San Antonio, Texas, 1954), pp. 80, 104. An example of how this myth is accepted in the academic community is Curtis, Sara Kay, *A History of Gillespie County, Texas, 1846-1900*, (M. A. thesis, University of Texas, Austin, Texas, 1943), p. 57.

2. Webb, Walter Prescott, *et al*, *The Handbook of Texas; A Dictionary of Essential Information* in Two Volumes, (Texas State Historical Association, Austin, Texas, 1952), p. 588; Tyler, Ron (ed.) *et al*, *The New Handbook of Texas* in Six Volumes, (Texas State Historical Association, Austin, Texas), Volume 5 p. 958; Buenger, Walter L., *Secession and the Union in Texas*, (University of Texas Press, Austin, Texas), p. 174; Richardson Rupert N., *et al*, *Texas*, (Prentice-Hall, Englewood Cliffs, New Jersey, 1981), pp. 222, 226; Baum, Dale, *The Shattering of Texas Unionism: Politics in the Lone Star State During the Civil War Era* (Louisiana State University Press, Baton Rouge, Louisiana, 1998), p. 242; and *Austin Texas State Gazette*, February 23, 1861. None of these references agree on the numbers voting for or against secession.

3. Ibid. As stated none of the above references agree on the Secession Ordinance. The author has used the numbers of Professor Baum who has made one of the most comprehensive studies of the Civil War era voting.

4. Baum, Dale, *Shattering of Texas Unionism*, Chapter 2.

3 – No Protection on the Frontier?

MYTH: After secession and the withdrawal of Federal forces the Texas frontier had "no protection at all. In a word, the settlements were left wide open to thieves, desperados, and all kinds of Indian depredation." [1]

FACT: After secession the Texas frontier had just as good, if not better, protection than it had before. [2]

DISCUSSION: At the beginning of the war, the Federal government had most of four regiments stationed in Texas. This was somewhere between ten and fifteen percent of the US Army; some thirty-seven companies or about 2,166 troops. The problem was the majority of this force was infantry and unsuited for mounted warfare against the Indians. Of these thirty-seven companies only ten were cavalry and suitable for offensive operations. [3]

At the start of the war or shortly thereafter the Texas Legislature established several types of part-time units for protection. The first was each of the frontier counties were authorized to raise 'Home Guard' or 'Minutemen' companies of 40-men each. [4] In Gillespie County the company was commanded by Philip Braubach, the sheriff and a radical insurgent. [5] In Kerr County the company was commanded by William T. Harbour. [6] In Blanco County the company was commanded by William T. Blackwell. [7] Almost all the members of Braubach's Company were Unionist, many later shown as insurgents. About a third of Harbour's company was Unionist, many later shown to be insurgents. About a third of Blackwell's company was Unionist, many later revealed to be insurgents. [8] The Home Guard Companies were disbanded a year later. [9]

In addition to the home guard units, two Confederate units were raised for frontier protection. These were the 1st and 2nd Regiments, Texas Mounted Rifles. [10] For the first year of the War, the 1st and 2nd Regiments were deployed along Texas' western frontier. The 1st Regiment was stationed along the Texas frontier from the Red River to Fort Mason, near the present-day town of Mason. The 1st Regiment also garrisoned Fort McKavett, roughly eighty miles northwest of Fredericksburg in what is now Menard County.

The 2nd Regiment was deployed along the Rio Grande from Fort Brown to Fort Duncan, at Eagle Pass. It had companies and detachments in many of the abandoned Federal forts and camps. Forts Inge, in Uvalde County, and Clark, in present Kinney County, Camp Wood, in present Real County, Hudson in present Val Verde County, and Verde, in Kerr County were among those garrisoned. [11]

By late 1861 it became clear to Texas that the two regiments providing frontier protection were going to be withdrawn for service in the east. There remained Confederate units at Camp Mason and Camp Verde. In response to this new development, Texas reorganized forces for both frontier protection and in the case of Union invasion by creating two new types of state forces. First, was a full-time regiment designed for frontier protection. [12] Second, was a total reorganization of its militia system to prepare the state to meet an expected Union invasion. [13]

The new state regiment was the Frontier Regiment. It had ten companies of one hundred men. Nine companies were recruited from the area it was to protect, and one company from the remainder of the state. Two companies were stationed in the Texas Hill county. One company was made up of men from Gillespie, Hays, and Kerr Counties. It established two camps; one on White Oak Creek, west of Fredericksburg, and the other camp at Camp Llano, on the Llano River northwest of Fredericksburg. The second company was made up of men from Bandera, Blanco, Medina, and Uvalde Counties. It had a camp at Camp Verde, in south Kerr County. The other camp was on the Seco Creek in western Bandera County. It also had a small detachment at Camp Rio Frio in what is now eastern Real County. [14]

The new militia law placed each county in a brigade district. All free white males between the age of eighteen and fifty were required to enroll in one of the new militia if they were not in the Confederate Army or the Frontier Regiment. The Hill Country was in the 31st Brigade District. Gillespie County had an entire regiment, the 2nd, composed of two battalions of three companies each, for a total six companies; four of which were commanded by Unionist. Kendall County had three companies, two of which were commanded by Unionist. Bandera, Blanco, and Kerr Counties each had one company. Medina County had an independent battalion of four companies; at least three commanded by Unionists. [15]

ENDNOTES - Myth # 3

1. Castro Colonies Heritage Association, Inc., *History of Medina County, Texas* (Castro Colonies Heritage Association, Inc. Castroville, Texas, undated), pp. 9-10.

2. Smith, David Paul, *Frontier Defense in the Civil War: Texas's Rangers and Rebels*, (Texas A & M University Press, College Station, Texas) 1992 and oral interview by the author with David Paul Smith, April 6, 1997.

3. Analysis of strength and types of Federal Forces in Texas by the author.

4. An Act to Provide for the Protection of the Frontier of the State of Texas, passed by the Texas Legislature on February 7, 1861, Laws of Texas, 1822-1897, (Compiled by H. P. N Gammel, Gammel Book Company, Austin, Texas, 1898), p. 346.

5. Muster Rolls, Captain Philip Braubach's Company, February 7, 1861, May 25, 1861, August 25, 1861, November 25, 1861, and February 25, 1862, Texas State Archives, Austin, Texas.

6. Muster Roll, Captain William T. Harbour's Company, March 5, 1861, Texas State Archives, Austin, Texas.

7. Muster Roll, Captain William A. Blackwell's Company, May 4, 1861, Texas State Archives, Austin, Texas.

8. Analysis of Braubach, Harbour, and Blackwell muster rolls.

9. An Act to Provide For the Protection of the Frontier of the State of Texas, passed by the Texas Legislature on December 21, 1861, Laws of Texas, 1822-1897, (Compiled by H. P. N. Gammel, Gammel Book Company, Austin, Texas, 1898), pp. 452-454.

10. For disposition of the 1st and 2nd Regiments Texas Mounted Rifles, see General Order No. 8, Headquarters, Troops in Texas, San Antonio, May 24, 1861, *Austin State Gazette*, June 8, 1861, and Smith, *Frontier Defense in the Civil War*, p. 190.

11. An Act to Provide for the Protection of the Frontier of the State of Texas, passed by the Texas Legislature on December 21, 1861, Laws of Texas, 1822-1897, (Compiled by H. P. N. Gammel, Gammel Book Company, Austin Texas ,1898), pp. 452-454.

12. An Act to Perfect the Organization of State Troops, and Place the Same on a War Footing, passed by the Texas Legislature on December 25, 1861, Laws of Texas 1822-1897, (Compiled by H. N. N. Gammel, Gammel Book Company, Austin, Texas), pp. 455-465.

13. General Order Number 1, Headquarters Frontier Regiment, Texas Rangers, Austin, February 1, 1862, Adjutant Generals Correspondent, Texas State Archives, Austin, Texas.

14. Records, 31st Brigade District, Texas State Troops, A.G.C., TSA, Austin, Texas and Quarterly Returns, 31st Brigade, July 1, 1862 & October 1, 1862, A.G.C., TSA, Austin, Texas.

15. An Act to Perfect the Organization of State troops and Place the Same on a War Footing, passed by the Texas Legislature on December 25, 1861, Laws of Texas 1822-1897, (Compiled by H. N. N. Gammel, Gammel Book Company, Austin, Texas), pp. 455-465; Records 31st Brigade District, A.G.C, TSA, Austin Texas, Quarterly Returns, 31st Brigade, July 1, 1862 & October 1, 1862, A.G.C., TSA, Austin, Texas and analysis of 31st Brigade District Records.

4 – Union Loyal League; Threat or Not?

MYTH: The organization known today as the Union Loyal League was not a threat to the State of Texas or the Confederate governments. [1]

THE FACT: The Union Loyal League was the political arm of a well-organized pro-Union insurgency in the Texas Hill Country.

DISCUSSION: Sansom's 1905 pamphlet is one of two supposedly 'primary' sources which give the purpose of the League. But Sansom was not a member of the League and had no first-hand knowledge about their purpose. [2] The second is an interview with August Hoffmann, who was a member of the League. In the Hoffmann interview the interviewer added much information she had read from other accounts, such as the Henry Schwethelm account. A great deal of the added information was not what Hoffmann actually said. [3] All other books, articles, and accounts accept these statements as facts. [4] The League was in direct opposition to the State of Texas and the Confederate governments. By the summer of 1861 both had passed laws and issued proclamations against such organizations. [5] At the time the League was organized there was no action on the part of either the State or Confederate governments to disturb or force anyone to bear arms against the Union. The Confederate draft did not take place until April 1862, ten months after the League was organized.

The term 'Union Loyal League' is used for the first time in Sansom's 1905 Pamphlet. Curiously, at no time did the actual members of the League use this term. They simply referred to it as 'The Organization' or in similar terms. [6]

A small cell of Freethinkers and Forty-Eighters established the so-called Organization. This was a highly secretive group. [7] At no time were the State of Texas or Confederates aware that such an organization existed. The Freethinkers and Forty-Eighters use the model described today by political and military scientists as an insurgency. The organization of a secret political element is part of Phase I of a three-phase insurgency. They gained such organizational knowledge during their failed 1833 and 1848 Revolutions in Germany. They refined their knowledge once in Texas when they attempted to organize the Texas Germans into a political force in 1854. [8]

August Siemering, an original member of the League, in an 1876 article describes the purpose of the League. He says, "The result of this meeting was a secret (emphasis added) alliance that bound the participants, not only to aid each other, but to pledge their fealty to the United States. Secret signs, signals, passwords, and everything pertaining to a fraternal organization were determined upon . . ."[9]

The Organization known today as the Union Loyal League was part of a statewide underground movement of Unionist groups called 'Loyal Leagues.' The goal of these Loyal Leagues was to restore the Federal government to power. They were secret organization: with secret signs and oaths. The Organization, like other Loyal Leagues also had secret signs and oaths binding members together; if anyone broke this oath he was to be executed. The Organization executed at least one man, Basil Steward, because he broke his oath. [10]

The most damning evidence that the Union Loyal League was a threat to the State and Confederate Governments is what the insurgents themselves said the purpose of the league was. Ernest Cramer says the Organization's military element "had been formed of men gathered together with, the understanding that as soon as the Northern troops would come within reaching, we would join them." [11] August Siemering says of the Organization's military element, "Immediately plans were made to organize this company from Union people in the Gillespie, Kerr, and Kendall counties . . . " He goes on to say, "Other plans developed, call for this company to be the base for other organizations, which then, at the right moment can act on behalf of Union." [12]

ENDNOTES - Myth #4

1. Sansom, *Battle of Nueces*, p. 2.

2. Ibid.

3. Interview with August Hoffmann contained in newspaper article entitled, "The Blackest Crime in Texas Warfare" by Helen Raley in the *Dallas Morning News*, May 5, 1929.

4. Some of the later publications that use Sansom and Hoffmann as the source are: Fehrenbach, T. R., *Lone Star: A History of Texas,* (American Legacy Press, New York, New York, 1983), p. 363; Ransleben, Guido E., *A Hundred Years of Comfort*, p. 105; Biggers, D. H., *German Pioneers,* pp. 57-58; Glenn, Frankie Davis, *Capt'n John: Story of a Texas Ranger* (Nortex Press, San Antonio, 1991), p. 2; Wooster, Ralph A., *Texas and Texans in the Civil War*, (Eaken Press, Waco, Texas), p. 114; Glenn, Frankie Davis, *John William Sansom's Battle of the Nueces*, (Published by Frankie Davis Glenn, Boerne, Texas 1991), pp. 5-6; McGowen, Stanley S., *Horse Sweat and Powder Smoke: The First Texas Cavalry in the Civil War*, (Texas A & M University Press, College Station, Texas, 1999), p. 66; Smith, David Paul, *Frontier Defense in the Civil War*, (Texas A & M University Press, College Station, Texas, 1992), p. 156; Shook, Robert W., "The Battle of the Nueces, August 10, 1862", *Southwestern Historical Quarterly*, Volume LXV, October 1961, p. 32; Rutherford, Philip, "Defying The State of Texas", *Civil War Times Illustrated*, Volume 19, No. 1, April 1979, p. 17; Clare, Mary, "Bloody Ground: The Incident on the Nueces", *Civil War*, Issue Number 70, October 1998, p. 49; Kelton, Elmer, "The Fleeing Sixty-A True Story", *Ranch Romances Magazine*, January 18, 1952; Alberthal, Vernel I., "Bushwhackers In Them Thar Hills", *The Radio Post*, October 2, 1952; Schmidt, Eduard, English Translation of Address Commemorating The 50th Anniversary Of The Battle On The Nueces, August 1862, copy provided by Gregory J. Krauter, Comfort, Texas; Gold, Gerald R., "Gillespie County in the Civil War", *The Junior Historian*, date not known, p. 27; Felger, Robert Pattison, *Texas In The War For Southern Independence 1861-1865*, (Ph. D. Dissertation, University of Texas, Austin, Texas, 1947), p. 342; Curtis, Sara Kay, *History of Gillespie County*, ((M.A. thesis, University of Texas, 1943),

pp. 43-44, 57; Hall, Ada Maria, *The Texas Germans in State and National Politics, 1850- 1865*, (M. A. Thesis, University of Texas, Austin, Texas, 1938), p. 87; Weinheimer, Ophelia Nielsen, *The Early History of Gillespie County*, (M. A. thesis, Southwest Texas State Teachers College San Marcos, Texas, 1952), p. 56; Heintzen, Frank W., *Fredericksburg, Texas, During The Civil War And Reconstruction*, (M. A. Thesis, St. Mary's University, San Antonio, Texas; 1944), p. 342; Comfort Heritage Foundation Handout, Comfort, Texas; Baulch, Joe, "The Dogs of War Unleashed: The Devil Concealed in Men Unchained", *West Texas Historical Association Year Book*, Volume LXXIII, 1997, p. 130; and Dykes-Hoffmann, Judith. *Treue Der Union: German Texan Women On The Civil War Homefront*, (M. S. Thesis, Southwest Texas State University, San Marcos, Texas, 1996), p. 68.

5. Texas Governor's Proclamation Book, TSA, Austin, Texas; *San Antonio Herald,* June 8, 1861, 'An Act Respecting Alien Enemies', approved August 8, 1861, OR, Series II, Volume II, 1368-1369; 'An Act to Alter and Amend an Act Entitled "An Act For The Sequestration of the Estates, Property, and Effects of Alien Enemies and for Indemnity of Citizens of the Confederate States, and Persons Aiding the Same in the War With The United States', August 13, 1861, OR, Series II, Volume II, pp. 932-944 and Proclamation by Jefferson Davis, President Confederate States of America, August 14, 1861, OR Series II, Volume II, pp. 1368-1369.

6. Samson, *Battle of Nueces*, p. 2. Examples of how the Germans referred to the league as 'the Organization' are Siemering's *The Germans in Texas During the Civil War*, May 29, 1929: Letter from Ernst Cramer to 'My Beloved parents', Monterrey, Mexico, October 30, 1862, with an English translated copy provided to author by Gregory Krauter and Letter from Fritz Tegener, Austin, Texas to Herr August Duecker, Gillespie County, Texas, August 23, 1875, with an English translated copy provided to author by Gregory Krauter.

7. Siemering, August, *Ein Verstehl Tex Leben*, (Published by August Siemering, San Antonio, Texas 1876); Francis, Mary E., *The Hermit of the Cavern*, (Naylor Printing Company, San Antonio, Texas, 1932), p. 106; and Siemering, August, *The Germans During Civil War*, (Freie Presse fur Texas, May 29, 1923).

8. Biesele, Rudolph L., *The History of the German Settlements in Texas 1831-1861*, (Von Boeckmann-Jones Company, Austin, Texas, 1930), pp. 198-203. Also see Biesele, "The Texas State Convention of Germans in 1854", *Southwestern Historical Quarterly*, Volume XXXIII, April 1930.

9. Siemering, A., *Germans During the Civil War*, May 29, 1923.

10. Elliott, Claude, "Union Sentiment in Texas 1861-1865", *Southwestern Historical Quarterly*, Volume L, January 1947; Letter, Ernst Cramer to 'My Beloved parents', Monterrey, Mexico, October 30, 1862; Siemering, *Germans During Civil War*, June 3, 1923, and Weber, Adolf Paul, *Deutsche Pioniere, Zur Geschichte Des Deutschthums in Texas*, Selbstverlag Des Verfasserrs. 1894, pp. 12-13.

11. Letter, Ernst Cramer to 'My Beloved parents' Monterrey, Mexico, October 30, 1862. An English translated copy provided to author by Gregory Krauter of Comfort, Texas.

12. Siemering, *Germans During Civil War*, May 29, 1923.

5 – Only A Local Militia?

MYTH: The League's military battalion was just the local militia, designed to protect the citizens from Indian attacks and to keep the locals "toned down in case of agitation," organized solely as to protect the settlers from Indian attacks and to take such actions as might peaceably secure its members and their families from being disturbed and compelled to bear arms against the Union. [1]

FACT: Some portions of the myth have a basis in fact; that is, "to take such actions as might peaceably secure its members and their families from being disturbed and compelled to bear arms against the Union." The secret military battalion <u>was</u> fully prepared to use force to secure its members and their families from being disturbed and compelled to bear arms against the Union. The part about keeping the locals "toned down in case of agitation" is also partly correct. But the locals it spoke of were those citizens who were opposed to the Union.

DISCUSSION: As pointed out in the discussion of Myth #3, Texas created several types of militia units at the start of the war. First were the minutemen companies in the frontier counties. [2] Gillespie County had one such company commanded by Philip Braubach—a radical Unionist. [3] Kerr County had another, commanded by William T. Harbour. [4] Blanco County had such a company commanded by William T. Blackwell. [5] Bandera County had an unofficial company which was Unionist. [6] The exact type of militia unit Medina County had is not known. All these minutemen companies existed from February of 1861, until February, 1862. [7]

At the expiration of the term of the minutemen companies, the State of Texas organized two types of state troops; first, a full-time regiment stationed on the frontier. [8] Secondly, every county in the state was to organize militia units. Companies were created in each precinct of the county. If there were not sufficient men in one precinct for a company to be formed, then precincts were combined to form a company. [9] Counties were placed in a brigade district. The counties of the Hill Country were placed in the 31st Brigade District, headquartered at New Braunfels in Comal County Because of its population density, Gillespie County was authorized a regiment consisting of two battalions, each with three companies for a total of six companies. Bandera, Blanco, Kendall and Kerr Counties were combined into a second regiment. Kendall County had three companies. Bandera, Blanco, and Kerr Counties each had one company. Medina County had an independent battalion of four companies. [10]

It is this second type of militia that Sansom and others claim the Organization's military battalion was; just a "local militia designed to protect the citizens from Indian attacks." [11] Almost all writers and historians state this as a fact. [12] The Gillespie County Regiment, officially the 2nd Regiment, 31st Brigade District, consisted of two battalions each with three companies. [13] The regimental commander was State Colonel Charles Kothe. The First Battalion was commanded by State Lieutenant Colonel Charles Weyrick. The companies were Company A, commanded by insurgent State Captain Rudolph Radeleff; Company B, commanded by State First Lieutenant Jacob Schmidt; and Company C, commanded by State Captain Valentine Hohmann—a radical Unionist, whose son or nephew was in the August insurgent group. [14] The Second

Battalion was commanded by State Major Jacob Luckenbach, a radical Unionist, whose brother, August, was in the August insurgent group. The companies were Company D, commanded by a Unionist, State First Lieutenant Jacob Dearing; Company E, commanded by State Captain Jacob Kuechler, who was also the commander of the Organization's Gillespie County Company. Company F, was commanded by State Captain William Feller – also a radical Unionist. [15]

The Bandera, Blanco, Kendall and Kerr Counties were in the Second Regiment, officially the 3rd Regiment, 31st Brigade District, which also consisted of two battalions each with three companies. Surprisingly it was in the 3rd Regiment where Unionists had the greatest percent in command. The regiment commanded was State Colonel Frederick Tegener, the commander of the Organization's battalion. The First Battalion was commanded by State Lieutenant Colonel Julius Schlickum, an 1854 radical. Company A (Blanco County) was commanded by State Captain A. J. Kercheville. Company B (Kendall County) was commanded by State First Lieutenant Frederick Lenz. Company C (Kendall County) was commanded by State Captain Ottomar Labhardt, a Unionist. The Second Battalion was commanded by insurgent State Captain Ernest Cramer, who was also the commander of the Organization's Kendall Company. Company D (Kendall County) was commanded by State Captain Michael Lindner, likely a Unionist. Company E (Kerr County) was commanded by State Captain Thomas Saner, also likely a Unionist. Company F (Bandera County) was commanded by State Captain Braden Mitchell – again a likely Unionist. [16]

While it is understandable that this information might be confusing to a nonmilitary individual, it was necessary to show the number of militia companies in the five primary counties of the Organization's area of influence. To recap; there were two full-time Texas State companies in the Organization's area. These were dedicated to protect the citizens from Indian attacks. Added to this were the twelve militia companies in the 31st Brigade District which could be called to duty if there were Indian dangers. Of these twelve militia companies at least four were commanded by known Unionist. It is hard to comprehend that they believe that three Organization companies were to offer any additional protection. All the men in the league's battalion were already members of a militia company.

This evidence supports a contention that the Organization's military battalion was an organized unit positioned to oppose the State and Confederate governments should the Union invade.

ENDNOTES - Myth #5

1. Sansom, *Battle of Nueces*, p. 2.

2. An Act to Provide for the Protection of the frontier of the State of Texas, passed by the Texas Legislature on, February 7, 1861, Laws of Texas, 1822-1897, (Compiled by H. P. N Gammel, Gammel Book Company, Austin, Texas, 1898), p. 346.

3. Muster Rolls, Captain Philip Braubach's Company, February 7, 1861, May 25, 1861, August 25, 1861, November 25, 1861, and February 25, 1862, Texas State Archives, Austin, Texas.

4. Muster Roll, Captain William T. Harbour's Company March 5, 1861, Texas State Archives, Austin, Texas.

5. Muster Roll, Captain William A. Blackwell's Company, May 4, 1861, Texas State Archives, Austin, Texas.

6. Letter, Charles Montague, Bandera Justice of the Peace, Bandera, Texas, July 19, 1861, to Governor Clark, Texas State Archives, Austin, Texas.

7. An Act to Perfect the Organization of State Troops and Place the Same on a War Footing, passed by the Texas Legislature on December 25, 1861, Laws of Texas, 1822-1897, (Compiled by H. P. N. Gammel, Gammel Book Company, Austin, Texas, 1898), pp. 455-465.

8. Ibid., pp. 452-454.

9. Ibid., pp. 455-465.

10. Ibid.

11. Sansom, *Battle of Nueces*, p. 2.

12. See Chapter 4, Endnote #4, pp. 21-22 for a partial list of articles and books that contained this myth. Almost everything published prior to 1997 states this as a fact.

13. An Act to Perfect the Organization of State Troops and Place the Same on a War Footing, passed by the Texas Legislature on December 25, 1861, Laws of Texas, 1822-1897, (Compiled by H. P. N. Gammel, Gammel Book Company, Austin, Texas, 1898), pp. 455-465.

14. Report, Headquarters 31st Brigade, Texas State Troops, New Braunfels, Comal County, March 8, 1862, A.G.C., TSA, Austin, Texas.

15. Ibid.

16. Ibid.

6 – In Reaction to Duff's Partisan Rangers?

MYTH: The League organized its military battalion on July 4, 1862, in reaction to Duff's first visit to the Hill Country. [1]

FACT: The Organization's military battalion was organized on March 24, 1862, as a result of failure to gain control of the local company of the Frontier Regiment – a full-time state unit operating in Gillespie, Kerr, and Llano Counties. [2]

DISCUSSION: As previously noted, at the start of the War or shortly thereafter, the Texas Legislature established several types of part-time units for protection. The first was that each of the frontier counties were authorized to raise a 'Home Guard' or 'Minutemen' company of 40 men each. [3] In Gillespie County the company was commanded by Philip Braubach, county sheriff and a radical Unionist. [4] In Kerr County the company commander was William T. Harbour. [5] In Blanco County the company was commanded by William T. Blackwell. [6] Almost all the members of Braubach's company and about a third each of Harbour's and Blackwell's companies were Unionists, many later also shown to be insurgents. [7] The Home Guard companies were disbanded a year later in February, 1862. [8] The Organization fully controlled Braubach's Gillespie Company and used it to intimidate the citizens of Fredericksburg and Gillespie County. In speaking of anyone who did not support the Unionists, Captain Braubach threatened retaliation against those who signed a petition, saying, "He would bring two hundred men to their doors which would make them talk different." [9]

There was also another type of unit; those which had been authorized by an earlier militia law. Captain Charles Nimitz of Gillespie County, and Captain George Freeman's Pedernales Cavalry Company in Blanco County were two of the Hill Country units. [10] Most members of these two companies were not Unionists; most of Nimitz's company was anti-Unionist. [11]

A further type of local militia units were those ordered established by the Texas Legislature. Their mission was to protect the state in case of Union invasion. All free white males between the ages of eighteen and fifty were required to be enrolled one of these militia units. The law created thirty-three brigade districts. [12] The 'authorized' militia units in the Hill Country were the 31st Brigade District, headquartered at New Braunfels in Comal County. Bandera, Blanco, and Kerr Counties had one company each. Gillespie County had six companies. Kendall County had three companies. Medina County had four companies organized into an independent battalion. [13]

In late 1861 it became obvious to the Texas government that the two full-time Confederate regiments that had been provided as frontier protection were going to be transferred east. The Texas Legislature authorized a full-time state regiment dedicated to frontier protection. The Frontier Regiment consisted of ten companies and began organizing in early 1862. Most of the men eligible to enroll had to be from the frontier counties. [14] One company would be made up of citizens of Gillespie, Hays, and Kerr Counties. Another was composed of citizens of Bandera, Blanco, and Medina Counties. [15] These two full-time companies would be stationed in the insurgents' area of influence. The insurgents felt they must control the two companies, or they would be a threat to them. Their attempt failed completely in Bandera, Blanco, and Medina Counties, but in Gillespie County, the individual authorized to raise the company was Jacob Kuechler, another radical insurgent. He enlisted only insurgents and Unionists. Area citizens complained [16] to the governor and as a result Kuechler's Company was ordered to disband. In the effort to organize a replacement company, most of Kuechler's men attempted to enroll. But Kuechler was denied enrollment and the other insurgents "left in a lump." [17]

The insurgents now believed, "Due to the aggressive politics of the secessionist, now saw it as their duty to organize themselves and in case of an emergency (or act of aggression) to resist." [18] The insurgents sent messengers to all the nearby counties to call supporters of the Union to a meeting. Of course the meeting had to be kept secret and a place in the mountains, where Bear Creek originates on the watershed between Fredericksburg and Comfort was selected. [19] On March 24, 1862, representatives from Comal, Gillespie, Kendall, Kerr, and Mason attended. The military battalion was organized by counties, each commanded by a captain. Fritz Tegener of Kerr County was elected major the commander. [20] Companies were organized by counties. Gillespie Company was commanded by Jacob Kuechler, and Kendall Company by Ernst Cramer. A third company of Anglos, called the American Company was commanded by Henry Hartman. [21] No evidence has been located showing that Comal and Mason counties ever organized a company.

The league's (the Organization) military battalion was organized on March 24, 1862, as a result of the loss of Braubach's minuteman company and the failure to control the full-time Frontier Regiment Company – not a reaction to Duff's first visit.

ENDNOTES - Myth #6

1. Samson, *Battle of the Nueces*, p. 2.

2. Cramer's Letter and Letter Julius Schlickum, dated December 21, 1862 on board the English frigate *Hope*, December 21, 1862, to his father-in-law.

3. An Act to Provide for the Protection of the Frontier of the State of Texas, passed by the Texas Legislature on, February 7, 1861, Laws of Texas, 1822-1897, (Compiled by H. P. N Gammel, Gammel Book Company, Austin, Texas, 1898), p. 346.

4. Muster Rolls, Captain Philip Braubach's Company, February 7, 1861, Mary 25, 1861, August 25, 1861, November 25, 1861, and February 25, 1862, Texas State Archives, Austin, Texas.

5. Muster Roll, Captain William T. Harbour's Company, March 5, 1861, Texas State Archives, Austin, Texas.

6. Muster Roll, Captain William A. Blackwell's Company, May 4, 1861, Texas State Archives, Austin, Texas.

7. Analysis of Braubach, Harbour, and Blackwell muster rolls.

8. An Act to Provide For the Protection of the Frontier of the State of Texas, passed by the Texas Legislature on December 21, 1861, Laws of Texas, 1822-1897, (Compiled by H. P. N. Gammel, Gammel Book Company, Austin, Texas 1898), pp. 452-454.

9. "Records of the Confederate Military Commission in San Antonio, July 2-October, 1861" Edited by Alwyn Barr in *Southwestern Historical Quarterly*, Volume LXXI, No. 2, October 1967, p. 263.

10. Muster Roll, Captain George Freeman's Company, March 1, 1861, & November 18, 1861, Texas State Archives, Austin, Texas and Muster Roll, Captain Charles H. Nimitz's Company, July 31, 1861, Texas State Archives, Austin, Texas.

11. Analysis of Nimitz and Freeman muster rolls.

12. An Act to Perfect the Organization of State Troops and Place the Same on a War Footing, passed by the Texas Legislature on December 25, 1861, Laws of Texas, 1822-1897, (Compiled by H. P. N. Gammel, Gammel Book Company, Austin, Texas, 1898), pp. 452-454.

13. Report, Headquarters 31st Brigade, Texas State Troops, New Braunfels, Comal County, March 8, 1862, A.G.C., TSA, Austin, Texas.

14. An Act to Provide For the Protection of the Frontier of the State of Texas, passed by the Texas Legislature on December 21, 1861, Laws of Texas, 1822-1897, (Compiled by H. P. N. Gammel, Gammel Book Company, Austin, Texas, 1898), pp. 452-454.

15. Ibid.

16. Letter, D. H Farr, Kerrsville, Kerr County, February 13, 1862, to Governor Lubbock, Governor Lubbock's File, Texas State Archives (TSA), Austin, Texas: Letter from Frank V. D. Stucken, Fredericksburg, February 13, 1862, to Governor Lubbock, Governor Lubbock's File, TSA, Austin, Texas: Petition, Citizens of Kerr County to Governor Lubbock, dated February 14, 1862, Governor Lubbock's File, TSA, Austin, Texas: and Siemering, *Germans During Civil War*, May 29, 1923.

17. "The Diary of D. P. Hopkins", *San Antonio Express*, January 13, 1918, and Siemering's *Germans in the Civil War*, May 29, 1923.

18. Siemering, *Germans in the Civil War*, May 29, 1923.

19. Ibid.

20. Letter, Ernst Cramer to 'My Beloved parents' Monterrey, Mexico, October 30, 1862. An English translated copy provided to author by Gregory Krauter of Comfort, Texas.

21. Sansom, *Battle of Nueces*, p. 3.

7 – Gillespie Unionists Had Majority Support?

MYTH: The secession referendum vote demonstrated that the Union Loyal League and its military battalion were supported by a majority of the Gillespie citizens. [1]

THE FACT: A large segment of Gillespie Germans were opposed to the League.

DISCUSSION: Representative of the area Germans who opposed the League were Charles Nimitz and his Gillespie Rifles Company. On February 23, 1862, they passed a resolution condemning the activities of the League. The resolution stated in part, "who by their (the insurgent leaders) teaching have been, and are still openly and covertly demoralizing the people of said County, by endeavoring to demonstrate to them the weakness and instability of the Government of the Confederate States and in various other ways doing all in their power to organize encourage an opposition to the southern cause, who openly speak of their party as one opposed to southern rights and institutions, and whereas we deem the aforesaid men dangerous to our community. Therefore be it resolved that they be warned to desist from the course heretofore pursued by them, or else that measures will be adopted to prevent their doing further injury to our community." [2]

There were also somewhere between 28 and 74 Gillespie citizens who signed a petition opposing Kuechler's attempt to recruit a company for the Frontier Regiment. [3]

Not only did Nimitz's Gillespie Rifles condemn the League, but so did some of the early radical Forty-Eighters. One potential supporter explained how the belief that Union would soon raise its flag over Texas "prompted the (insurgents) to form and organize a sort of secret brotherhood." He was told the League was mainly for protection against surprise attacks, burning of the settlements and hangings perpetrated by the southern-oriented parties. But he understood the real purpose. He was approached to join and take command. "With fearful worry," commented the individual, "did I watch the unfolding of this association because I was too familiar with circumstances not to see the danger and futility of this undertaking. I knew too well that a few backwoods men could not lead war against the State of Texas: A State who had about 15,000 in service with weapons." Then he mused, "What did we have?" He answered his own question, "Hunting guns, little powder, and if we were to battle, no line of retreat. In spite of all rumors, I could not believe in an early arrival of the Union Army. The Government of the U. S. could not send troops to Texas, as long as the situation in Virginia and Tennessee was uncertain." He "emphatically" refused to be part of this undertaking and tried with all his power to reason and convince the League leaders into reconsidering this "foolish attempt." He repeatedly predicted to the League leaders the ruin of the settlement in the end. "In vain!" [4]

ENDNOTES - Myth #7

1. Tyler, Ron, (ed), *The New Handbook of Texas*, Volume 3, p. 166.

2. Minutes of Meeting, Gillespie County Rifles, February 23, 1862, and March 29, 1862, with copy of the Gillespie County Rifles Resolution, District Clerk's Office, Fredericksburg, Texas and Letter Julius Schlickum, dated December 21, 1862, to his father-in-law.

3. Elliott, Claude, "Union Sentiment in Texas 1861–1862", *Southwestern Historical Quarterly,* Volume L, January, 1947, footnote 45, p. 463, and Charles Nimitz's testimony, Barr, Alwyn, "Records of the Confederate Military Comission in San Antonio July 2-October 10, 1862", *Southwestern Historical Quarterly*, Vol. LXXI, October, 1967.

4. Letter, Julius Schlickum, to his father-in-law, dated December 21, 1862, on board the English frigate *Hope*.

8 – Innocent Neutrality?

MYTH: The members of the Union Loyal League were neutral in sentiment and were simply just innocent individuals who did not want to fight for ether side. [1]

FACT: The League (Organization) members were organized insurgents who intended to fight when the Union invaded Texas and to rise up and declare the western part of Texas (now the Texas Hill Country) as the Free State of West Texas. [2]

DISCUSSION: The League's (Organization's) military battalion was described in the words of one of its military commanders, "Our Company had been formed of men gathered together with the understanding that as soon as the Northern troops would come within reaching distance, we would join them." [3]

All of the accounts written by members of the fleeing insurgent group stated they would flee and join the Union Army. August Hoffmann said, "Sixty-five or sixty-eight men set out for the Rio Grande, intending to cross into Mexico whence they would sail for New Orleans and join the Union Army." [4] Ernest Cramer said, "We would go to Mexico where there might be a chance for us to join the Northern forces." [5] Jacob Kuechler said, "A party of about 62 Union men met in camp on Turtle Creek, in Kerr County, to leave Texas and join the Federal Army by way of Mexico." [6] August Siemering, an original member of the Union Loyal League, stated, "Finally the decision was made to forsake the land, cross the Rio Grande into Mexico and offer themselves to the United States Army." [7]

The first written account of the event was made public in 1866. It stated the insurgents planned, "to proceed to Mexico and, if possible, to join the Union Army." [8]

From their own words, it can be concluded that the fleeing group first hoped to join the Union Army when it invaded; and secondly, once they understood the Union Army was <u>not</u> going to invade, decided to flee to Mexico and on to New Orleans to enlist in the Union Army.

ENDNOTES - Myth #8

1. Fehrenbach, *Lone Star*, p. 363; Pirtle III, Caleb and Cusack Michael F., *Fort Clark, The Lonely Sentinel On Texas's Western Frontier,* (Eakin Press, Austin, Texas, 1985), p. 50; Lonn, Ella, *Foreigners In The Confederacy*, (The University of North Carolina Press, Chapel Hill, North Carolina, 1940), p. 426; Biggers, *German Pioneers in Texas*, p. 58; Michener, James. *Texas*, (Random House, New York, New York), p. 624; Gurasich, Marj, *A House Divided*, (Texas Christian University Press, Fort Worth, Texas, 1994), p. 33; Gold, Gerald R., "Gillespie County in the Civil War" in *The Junior Historian*, May 1, 1965, p. 30; Schmidt, Eduard, English translation of Address Commemorating the 50th Anniversary of the Battle on the Nueces, August 1862, copy provided by Gregory J. Krauter, Comfort, Texas; Kelton, Elmer, "The Fleeing Sixty-A True Story" *Ranch Romances Magazine,* January 18, 1852; Lossing, Benson J., *Pictorial Field Book of The Civil War*, 1997, Pp. 536-537; Baulch, Dogs of War Unleashed; Biffle, Kent, 'Remembering Hill Country Bad Old Days" *Dallas Morning News*, November 23, 1997; and Sansom, *Battle of Nueces*, pp. 3-4.

2. Cramer's Letter; Siemering, *Germans During Civil War*, May 29, 1923, and Barr CMC, Volume LXXI, October 1967, p. 264.

3. Cramer's Letter.

4. Raley, Blackest Crime in Texas Warfare.

5. Cramer's Letter.

6. Kuechler's Letter, as quoted by Guido Ransleben in *A Hundred Years of Comfort in Texas: A Centennial History*, (The Naylor Company, San Antonio, Texas, 1954), p. 96.

7. "German Unionists in Texas", *Harper's Weekly*, New York, New York, January 20, 1866.

9 – Confederate Overreaction?

MYTH: The Confederate authorities completely overreacted to the organization of the Union Loyal League and declared the Hill Country counties in open rebellion, declared martial law, and appointed James Duff provost marshal. [1]

FACT: There are different events incorporated in this myth. First was General Paul O. Hebert's statewide declaration of martial law on May 31, 1862. [2] Second, when in early August of 1862, General H. P. Bee sent Confederate troops to the Hill Country to "issue a proclamation declaring martial law, and required all good and loyal citizens to return quietly to their homes, and take the oath of allegiance to the Confederate and State governments, or be treated summarily as traitors in arms." [3]

DISCUSSION: General Herbert's statewide declaration of martial law had nothing to do with the organization of the Union Loyal League; the Confederates did not know of the existence of the League. The declaration was to deal with those who refused to accept Confederate paper money at par, and those who avoided the draft. [4]

General Bee sent several Confederate units to various locations to inform the citizens of the contents of martial law. One of these units was the Partisan Company commanded by James Duff which was sent to the Hill Country. [5] Duff's company arrived on May 30, 1862, and established camp at the old federal frontier defense fort, Fort Martin Scott. Duff was <u>not</u> appointed provost marshal at that time. The name of the Gillespie County provost marshal is not known, but it was likely William Wahrmund, Chief Justice or Charles Nimitz, who was later appointed the Confederate enrollment office. The first thing Duff did was, "immediately proclaimed martial law as existing within the limits of (Gillespie) County, (and) in Precinct No. 5 of Kerr County, giving six days to enable the citizens to report to the provost-marshal and take the oath of allegiance." [6] He followed the example used by General Bee in San Antonio. In essence, Duff rode his entire command of 100 men into town in a show of force and very dramatically read the proclamation of martial law. [7] This was followed by a citizen of Gillespie County, Fritz Messenger, reading the proclamation in German. Duff next posted men on different roads, streets, and lanes; and refused to let any person proceed without a pass. [8]

There are some negative comments that appeared in print about Duff's first tour at Fredericksburg. Probably the most damning was written by R. H. Williams, a member of Duff's company. Williams' entire narrative is very critical of Duff, but it was not published until 1907 by either a son or grandson of Williams. Williams says, "All we had done was to bully a few inoffensive Germans." [9] *The San Antonio Herald* in its June 11th issue reported that Duffs' company had been falsely accused of mistreating a woman in Fredericksburg. [10] However, other accounts tell a different story. The San Antonio Herald's article goes on to say, "No opposition had been shown to martial law and all was well in Fredericksburg." [11] To counter Williams' comment, another trooper in Duff's company said, "I am indeed very well pleased with the company, and all the men seem to be very kind and accommodating, all well-behaved. [12]

On their first trip to the Hill County Duff's company committed no atrocities. No farms or houses were burned. No one was executed. Perhaps the best conclusion of Duff's first visit to Fredericksburg in summed up in Duff's official report. In it he says, "I found the people shy and timid. I visited, with a part of my company, several of the settlements and explained to the people the object of our visit to their county. In a few days they displayed much more confidence in us, and in a corresponding ratio more desire to serve the Government." [13] Duff's company returned to San Antonio on June 21st.

In late July and early August, 1862, General Bee received request from several Gillespie County citizens asking for regular troops as they feared the activities of the insurgents. In early July the insurgents murdered Basil Stewart and attempted to kill a Confederate officer, Captain Van der Stucken. They raided Doss's mill and Gibson's farm in the western part of Gillespie County and assaulted one of the Basse brothers in Fredericksburg. Confederate authorities also received information telling of large depots of supplies. [14]

Captain Henry T. Davis, the commander of the company from the Frontier Regiment stationed in west Kerr County, reported the insurgents threatened an attack on smaller and less well-defended camps and that his scouts met them in the mountains, in groups numbering from twenty-five to forty men, and which his men could not take any action against. Davis expressed his fears that the insurgents would soon attack and capture their horses. He also requested regular Confederates be sent to the area. [15]

General Bee quickly responded to these requests. He ordered two task forces to the Hill County with orders to issue proclamations, to again declared martial law, require all good and loyal citizens to return quietly to their homes, and – for the third time – to take the oath of allegiance to the State and Confederate Governments, or be treated summarily as traitors in arms. He issued further orders to find and break up any insurgent encampments and depots and to send the families and provisions back to the settlements." [16]

The first task force was sent to the northern counties of Burnet, Llano, Mason, and San Saba. Lieutenant Colonel Nat Benton from the 32nd Regiment Texas Cavalry commanded this task force. Benton's task force consisted of elements of three companies from the 32nd Regiment Texas Cavalry with a total strength of about 220 men. Benton was both the provost marshal and troop commander. [17] The second task force was sent to the counties of Blanco, Gillespie, Kerr and Kendall. Unlike Benton's task force, the duties of provost marshal and troop commander were split. James Duff was the provost marshal, while Captain John Donelson was the troop commander. [18] The Donelson task force consisted of elements of four companies, to include that of James Duff although Duff was not in command of his company. The two task forces arrived in late July or very early August. Captain Duff established his provost marshal headquarters at Fort Martin Scott, while Captain Donelson established his headquarters at Camp Pedernales, on the Pedernales River about ten miles southwest of Fredericksburg. Benton's task force completed its mission by August 20, 1862, and returned to San Antonio. Benton reported all the men took the oath of allegiance and registered for the draft. [19]

This was not the case in Captain Donelson's area of responsibility. According to General Bee, many citizens refused to return to their homes. "Numerous small encampments with large supplies of provisions were found,"

Bee reported. He went on to say while "large numbers of young men returned to their homes, took the oath of allegiance, and enlisted in the army . . . it became certain that there were still many in arms who were determined to resist the (State and Confederate) Governments at all hazards." [20]

Donelson's task force planned to remain in the Hill County for only about six weeks. [21] Captain Donelson complied with General Bee's orders. He conducted what in current military term is called a counter-insurgency operation, which is primarily designed to separate the insurgents from their bases of supplies and personnel. In other words Donelson wanted to deprive the insurgency of food, weapons, and other useful materials, as well as to make recruitment from the population more difficult. [22]

The Confederates had previously learned the names of leading Unionists. It was these individuals that Donelson targeted. For example, Williams tells about farms, to include one owned by a man named Henderson, which were visited by Confederate troops and family members arrested as well as supplies returned to the settlements. [23] Another farm visited was that of Herman Nelson. [24] A third home which Confederate troops visited was Fritz Tegener's house and mill. [25] Fritz Tegener avoided being arrested, although the home of Tegener's brother, Gustav was also visited and he was arrested. [26] Donelson's troops also visited the home of Frank Scott who was arrested. Several family members of these men were detained. [27] The Christian Dietert mill in Kerr County was also visited. [28] In most of these cases insurgents or members of their families were arrested. There are several examples, such as that of Christian Dietert, released because of appeals from their families. [29]

There are no documented cases where any farms owned by non-Unionist where raided. Nor is there documentation to prove that 'suspected' Unionist farms were targeted by raids. It can certainly be concluded that Confederate troops visited farms in the Hill Country. However, there is no documentation to prove any arrests were made in these homes.

Donelson's actions were in compliance with General Bee's instructions. The result of the Donelson's task force was, as stated by General Bee, that his "instructions were fully carried out. Numerous small encampments with large supplies of provisions were found, far more than could possibly have been needed by those found in possession of them, chiefly women and children, who by their language and conduct removed all doubt, if any could still have existed, as to the purpose for which these supplies were intended. These were all removed to the settlements, or destroyed when the former course could not be pursued." [30]

ENDNOTES - MYTH #9

1. Sansom, John W., "The German Citizens Were Loyal To The Union", *Hunter's Magazine*, November 1911, Bandera, Texas; Glenn, Frankie David, *Capt'n John: Story of a Texas Ranger*, (Nortex Press, Austin, Texas, 1991), pp. 22-23.

2. General Order Number 45, Headquarters, Department of Texas, Houston, Texas, May 30, 1862, 'War of the Rebellion: A Compilation of the Official Records of the Union and Confederate Armies' (OR), Series I, Volume 9, pp. 715-716.

3. Report Brigadier General H. P. Bee, Headquarters Sub-Military District of the Rio Grande, San Antonio, Texas October 21, 1862, to Headquarters First District of Texas, San Antonio, Texas OR, Series I, Volume LIII, pp. 454-455.

4. Report, Brigadier General, P. O. Hebert, Headquarters First District of Texas, San Antonio, October 11, 1862 to General Cooper, Adjutant-General, Richmond, Virginia OR, Series I, Volume LII, pp. 828-829.

5. *San Antonio Herald*, May 31, 1862.

6. Report, James Duff, Captain, Commanding Company of Partisan Ranger, Headquarters Camp Bee, San Antonio, Texas, June 23, 1862 OR, Series I, Volume 9, pp. 785-787.

7. Smith, Thomas C., *Here's Yer Mule-The Diary of Thomas C. Smith, 3rd Sergeant, Company G, Wood's Regiment, 32nd (AKA as 36th Texas) Cavalry*, (The Little Texan Press, Wasco, Texas, 1958), pp. 11-12.

8. Betzer, Roy J., *Early Fredericksburg and Fort Martin Scott.* (unpublished MS, n. d., copy in possession of author), p. 4.

9. Williams, R. H., *With The Border Ruffians*, (University of Nebraska Press, Lincoln, Nebraska, 1982), p. 232.

10. *San Antonio Herald*, June 11, 1862.

11. Ibid.

12. Betzer, Roy J., *Early Fredericksburg and Fort Martin Scott*, (unpublished MS), n.d., p. 4, copy in possession of author); Duff's Report, p. 785.

13. Duff's Report, p. 785.

14. Report, Brigadier General H. P. Bee to Headquarters First District of Texas. October 21, 1862, OR, Series I, Volume LII, p.s 454-455.

15. Report, Captain H. T. Davis, to Headquarters Texas Frontier Regiment, July 25, 1862, A.G.C. TSA.

16. Bee's Report, pp. 454-455.

17. Lieutenant Colonel Nat Benton's Report, *San Antonio Weekly Herald*, August 30, 1862.

18. Bee's Report, pp. 454-455.

19. Benton's Report.

20. Bee's Report, pp. 454-455.

21. Williams, Ruffians, p. 235.

22. Thompson, Leroy, *The Counter Insurgency Manual*, (The Military Book Club, Stackpole Books, Mechanicsburg, Pennsylvania 2002), p. 83.

23. Williams, *Ruffians*, p. 238.

24. Duff's Repot, p. 786.

25. Siemering, *Germans During Civil War*, June 8, 1923

26. Bennett, Bob, *Kerr County Texas 1856 -1956*, (The Naylor Company, San Antonio, Texas 1956), p. 145.

27. Ibid.

28. Nixon Jr., Victor, "An Encounter With The Partisan Rangers", *The Junior Historian*, Texas State Historical Association, Austin, Texas, May 1, 1865, pp. 8-9.

29. Ibid.

30. Bee's Report, pp. 454-455.

10 – Duff's Company; Bullies and Renegades?

MYTH: Duff's Company was made up of bullies, outlaws, and cutthroats, some of whom had been let out of jail to join the Confederate Army. [1]

FACT: Duff's Company was made up of a cross-section of individuals, mainly from Bexar and surrounding areas of south and west Texas.

DISCUSSION: James Duff was a successful San Antonio merchant who had several contracts to supply the Federal forces in Texas with supplies. He had a net worth of over $35,000. [2] Duff's second in command was the former mayor of San Antonio. [3] His other two officers were prominent Bexar citizens. One was in business as a 'Lime Burner' with a worth in 1860 of over $17,000, [4] the other an attorney-at-law. [5] His first sergeant was a successful Boerne merchant, who later became a Confederate officer. [6] Three of his other non-commissioned officers were later Confederate officers. The troopers included thirteen farmers, twelve stock raisers, four clerks; three bakers, three lawyers, two laborers; two land agents; two ranchers; two who worked for stagecoach companies; two students; a wheel-wright, a tailor, a gardener, an alderman, a baker, a silversmith, a Baptist preacher, a sheep raiser, a bookkeeper, a gunsmith, a bar operator, a craftsman, a printer, a physician, a soldier and a saddletree maker. Not a single convicted criminal among them. [7]

Seventy-one troopers' occupations can be established; of twenty-nine, less than half worked the land in some way. Forty-two were 'city boys.' The men were generally of a higher social class than the average Texan recruit.

Duff's 91-man company was organized on May 7, 1862, as a direct result of the Confederate draft. [8] When Duff began recruiting he was still the militia brigadier general in command of the 30th Brigade District and thus was able to recruit outstanding individuals. All of his officers had been militia officers, most in command of companies. James R. Sweet was the first lieutenant of the Alamo Rifles: Richard Taylor was the first lieutenant of Company Number 3, Precinct Number 2; and Edwin Lilly a captain in command of a company in the 5th Independent Battalion. Four of his non-commissioned officers had been militia officers: George Horner was a captain in command of Company Number 3, Precinct Number 3; Emile Abat was a captain in command of Company Number 3, Precinct Number 2; George B. Pue was a captain in command of a company from precinct number 9 and 13; and George Caldwell was a first lieutenant in command of Company Number 3, Precinct Number 1. In addition several of his enlisted men had been militia officers. These included: Charles Hummel, a captain in command of Company Number 2, Precinct Number 2, Precinct Number 1; Edward Gallagham, a captain in command of a company from precinct number 22; Francis J. Mullins, a second lieutenant in Company Number 2, Precinct Number 2; all from the 30th Brigade District. C. C. Kelly had been a first lieutenant of a Kerr Militia Company. [9]

Since all Texas males between the ages of 18 and 50 were in the Texas State Militia it seems safe to assume that most members of Duff's Company had militia experience. [10]

Thirty-five men were born in southern states, fourteen were of foreign birth; thirteen in border states and seven in northern states These included nine born in Texas, nine in

Kentucky, five in Alabama, five in Tennessee, five in Germany, four in Mississippi, four in Georgia, three in New York, three in Virginia, two in Canada, two in Mexico, two in Scotland, two in Louisiana, two in Maryland, two in North Carolina; two in England; and one each in Alsace-Lorraine, Florida, Indiana, Missouri, New Jersey, Ohio, Pennsylvania, and the Cherokee Nation.

ENDNOTES - MYTH #10

1. Edwards, Walter F., (ed.), *The Story of Fredericksburg, Its Past, Present, Points of Interest and Annual Events*, (Fredericksburg Chamber of Commerce, Fredericksburg, Texas), n .d., p. 45; Edwards, Walter F., *Tales of Old Fredericksburg*, (Published by Walter F. Edwards, Fredericksburg, Texas), 1975, p. 2; and Handout The Comfort Heritage Foundation, Inc.) n. d, Comfort, Texas.

2. Eighth U. S. Census, 1860 Bexar Census, p. 3b.

11 – Captain Duff's Reign of Terror?

MYTH: Immediately upon arriving at Fredericksburg and the surrounding German counties Duff arrested many innocent prominent citizens. The burning of crops and homes, raping of women, and other various atrocities followed, all in the name of martial law. It is believed between 100 to 150 men and boys were lynched during Duff's tenure in the German counties. Duff conducted a reign of terror in the Hill County. [1]

FACT: As in Chapter 9, there are two separate elements in this myth. Captain Duff was sent twice to the Hill County. The first was in late May of 1862, to declare martial law and inform the citizens of martial law requests. [2] The second was in late July or early August, 1862, when he was appointed provost marshal of the area in support of Captain John Donelson's task force. [3] It is a fact that Unionists and insurgents felt like a "Reign of Terror" was being administered upon them.

DISCUSSION: In early June, 1862, Duff issued the martial law proclamation. As part of the proclamation, the citizens had six days to take the Confederate Oath of Allegiance. At first he found very few who would identify the Unionists. Captain Duff and Lieutenant Sweet issued summons of some of the more prominent citizens, to include Captain Charles Nimitz. "They obeyed the order and made affidavits in regard to certain citizen of the county" reported Duff. At the end of the six-day grace period Captain Duff began arresting Unionist/Insurgent leaders. These included: Philip Braubach, the sheriff; Friedrich Wilhelm Doebbler, a grocery keeper, and Heinrich Frederick Lochte, a merchant. Duff was unable to find and arrest militia Captain Jacob

Kuechler. [4] Several of those arrested were Unionists whom the Gillespie Rifles had denounced in February, 1862. The Gillespie Rifles referred to were those, "Who by their teaching have been, and are still openly and covertly demoralizing the people of said County, by endeavoring to demonstrate to them the weakness and instability of the Government of the Confederate States and in various other ways doing all in their power to organize and encourage an opposition to the southern cause, who openly speak of their party as one opposed to southern rights and institutions, and whereas we deem the aforesaid men dangerous to our community. Therefore be it resolved that they be warned to desist from the course heretofore pursued by them, or else that measures will be adopted to prevent their doing further injury to our community." [5]

On June 11th Captain Duff and his company left Gillespie County and moved to Blanco City, the county seat of Blanco County, where he declared martial law. Duff received a much warmer reception in Blanco County than he had in Gillespie County. In referring to Blanco County he said, "Here I found the great majority of the people friendly, enthusiastically so, to the Confederate States Government." [6] No arrests were made in Blanco County.

On June 19th Duff and his company arrived at Boerne, the Kendall County seat. As previously Duff proclaimed martial law and advised the citizens of the requirement. One suspected Unionist was arrested in Kendall County; [7] one of the 1854 ultra-radicals, Julius Schlickum. [8] Schlickum was also the commander of Company B, 3rd Regiment, 31st Brigade District. [9] As it turned out, Schlickum had refused to join the Union Loyal League. [10]

Duff's reasons for arresting Schlickum were because he "has been bitterly opposed to us and who I have reason to believe took an active part in forwarding expresses and information to Federal prisoners at Camp Verde and the disaffected citizens of his own and adjoining counties." [11] Captain Duff and his company left Boerne on the afternoon of the 20th and arrived back in San Antonio on June 21, 1862, "looking hale and hearty, but tolerable well dusted and somewhat sunburnt (sic), said one newspaper. [12]

Duff's company was in the Texas Hill Country from May 30th to June 20th, a total of twenty-two days. [13] All those arrested in Gillespie and Kendall Counties were convicted by the Confederate Military Commission of Unionist activities. [14] During his twenty-two days stay in the Hill County Captain Duff was very careful not to abuse his authority. No houses were burned, no one was killed, and no one was unfairly arrested. All those arrested were given 'due process' under martial law. The three Gillespie County men arrested were insurgents and key leaders of the league.

The second element of this myth lies in "histories" written by writers who have totally failed to tell the complete story. This in essence, is where the effort to 'demonize' Duff really begins. James Duff and his company returned to the Hill County in late July or early August of 1862. But this time, Duff did not command his company. It was under the command of Second Lieutenant Richard Taylor. [15] Instead, Duff was appointed provost marshal with his headquarters at old Fort Martin Scott, two miles east of Fredericksburg. [16] The troop commander was Captain John Donelson of the 2nd Regiment Texas Cavalry, at Camp Pedernales, ten miles west of Fredericksburg. [17]

To properly understand the missions given to Captain
Duff and Captain Donelson requires someone with a military
officer's background. General Bee's instructions should be
properly researched. Despite the hundred of published articles
on the subsequent events, no one references Bee's
instructions. This separation of authority is vital to
understanding the functions of James Duff as provost marshal
and Captain John Donelson as commander. Unless a historian
or writer can understand and interpret these relationships, they
are incapable of grasping what took place; many of the so-
called German experts who have written about the events and
personalities discussed here do not known what they are
talking about. The only way to fully comprehend what
happened is to scrutinize the orders under which Duff and
Donelson operated.

Nowhere in any of the previously-published accounts has
it been pointed out precisely what General Bee, the
Confederate commander, had instructed. General Bee's orders
stated, "I appointed Captain Duff provost-marshal for the
counties composing the disaffected district, and place under
his control (emphases added) four mounted companies,
commanded (emphasis added) by Capt. John Donelson. In
military terms control means the "authority that may be less
than full command." [18] Command means "the right of
authority to order control or dispose of subordinate
organizations. It also means the right to be obeyed." [19] The
state company in the area of Donelson's region – Company A,
Frontier Regiment – was not under the control or command of
the Confederate troops. [20]

General Bee clearly spelled out what each officer were to
accomplish. The provost marshal was to "issue a proclamation
declaring martial law, and requiring all good and loyal citizens

to return quietly to their homes, and take the oath of allegiance to the oath of allegiance to the Confederate and State governments." The military commander was "to send out scouting parties into the mountain districts with orders to find and break up any such encampment and depots, as had, been reported to exist there and to send the families and provisions back to the settlements." [21] If anyone refused to take the oaths of allegiance and to return to their homes they were to be "treated summarily as traitors in arms." [22] According to Webster summarily means "done without delay or formality; quickly executed." [23]

The provost marshal and troop commander even went to the pain of separating their physical locations. As provost marshal, Captain Duff established his office at old Fort Martin Scott. [24] Captain Donelson established his headquarters at Camp Pedernales, ten miles away. [25] A Post Return is the official record of the activities of a military camp of fort. The Camp Pedernales Post Returns clearly show what military units were there, who commanded them, and which officers were located at Camp Pedernales; the Post Return shows that Duff's Company was commanded by Second Lieutenant Richard Taylor. The company consisted of two officers, Lieutenants Taylor and Edwin Lilly, and 65 troopers. James Duff was not stationed at Camp Pedernales. [26]

The post returns also clearly tell of Donelson's actions. He, "with 100 men scouted the upper portion of Gillespie County for 5 days in search of certain bands of traitors to the C. S." Donelson arrested several Unionists when he discovered that a greater portion had left the county. He returned to Camp Pedernales and dispatched First Lieutenant Colin D. McRae with ninety-four fresh men and horses in pursuits of them." [27]

Let's address the myth again, which says that immediately upon arriving at Fredericksburg and the surrounding German counties Duff arrested many innocent prominent citizens. The burning of crops and homes, raping of women, and other various atrocities followed, all in the name of martial law. It is commonly believed that between one hundred and one hundred-fifth men and boys were lynched during Duff's tenure in the German counties. [28]

When Duff returned to Gillespie County the second time he arrested no one. He did not have the authority. That would have been accomplished by Captain Donelson. The August Post Returns proves this point when it says Donelson, "After arresting a great many . . ."[29] For the record, Donelson arrested no "innocent prominent citizens." Those arrested included known and suspected Unionists. Davis and Donelson released many of those who were suspected Unionists, including Christian Dietert, Robert Schaefer, and Adolph Rosenthal. [30] He confined only known Unionists and their families. Duff burned no homes. However, there is evidence showing Captain Davis' troops burnt some homes. [31] Duff raped no women. Nor does there exist any documented evidence that Davis' or Donelson's troops raped any women. They did make crude sexual comments to the wives held as prisoners. [32]

The statement that "between 100 to 150 men and boys were lynched during Duff's tenure in the German counties," is ludicrous, mainly because Duff was no longer in the Hill Country when the majority of Unionists were executed. Davis' and Donelson's men did execute several captured insurgents. This included five whom Donelson had arrested prior to the Nueces Battle. These five included a man named Howell (likely John Howell) who was hanged on August 4th near Camp Davis, where the full-time frontier company was stationed. [33] Davis' men were most likely the ones who executed Howell, although they denied it. [34] Donelson's men may have been the ones who executed Howell. This hanging took place at least fifteen miles west of Duff at Fort Martin Scott and some five miles west of Camp Pedernales. Captain Duff certainly had nothing to do with the deaths, as he had no troops to conduct a hanging. Donelson had also arrested four men about August 3rd near the Guadalupe River and Spring Creek; Sebird Henderson, Hiram Nelson, Gustav Tegener, and C. Frank Scott, who were all hanged near Spring Creek on August 22nd. [35] Two other insurgents, Heinrich Stieler and Theodore Bruckish, were executed about August 22nd on Goat Creek in northern Kerr County. The men who killed them were J. M. Seal and Alonze Rees of Davis' Company, Frontier Regiment. [36] Like the death of Howell, Captain Duff had nothing to do with these four deaths. All Confederate troops were under the command of Captain John Donelson and the state troops were commanded by Captain Henry Davis. Duff was no longer in the Hill Country when these Unionists were hanged on August 22nd. He and his company left on August 20th and returned to San Antonio not later than August 23rd. [37] However there is evidence that Duff may have been responsible for the death of two insurgents, Conrad Bock and Fritz Tays on August 24th near Boerne. A detachment from Duff's company captured these two and hanged them near Cibolo Creek, just north of Boerne. Duff and Taylor were indicted by a Kendall County Grand Jury. [38]

James Duff's second trip to the Hill Country lasted at the most twenty-one days. The earliest he could have arrived was on July 31. He left on August 20th and was back in San Antonio by August 23rd. [39] As pointed out in the preceding paragraph, by the time the majority were executed, Duff and his company had returned to San Antonio. To restate; not only was Duff not in the Hill Country at that time, neither was his company!

There were other insurgents arrested and executed, but again, not by James Duff. These insurgents were survivors of the Nueces Battle. Two were executed in the Hill Country, likely by Davis' men; William F. Boerner and Herman Flick. Both were executed about August 25th [40] – again, long after Duff and his company returned to San Antonio.

Three other Nueces Battle survivors were captured and executed. All three were captured west or northwest of San Antonio. The units who executed them are not known, but again it was not James Duff who was responsible for their deaths. The reason this can be stated so categorically is by the testament of R. H. Williams, who was a member of Duff's Company and had returned to it by the time these three was killed. Williams was extremely critical of Duff and would have included the executions in his book, if Duff had anything to do with them. Those executed were August Luckenbach, Adolph Ruebsamen, and his brother Louis Ruebsamen. [41]

To further accuse James Duff of being responsible for the deaths of several other Hill County citizens, there are several tombstones in the area claiming those buried in these were *getötet* (killed) by Duff. First are the tombstones of the four men hanged on Spring Creek on August 22nd; Sebird Henderson, Hiram Nelson, Gustav Tegener, and C. Frank Scott. Their epitaph reads "Hanged and Thrown in Spring Creek "By Col. James Duff's Confederate Regiment." [42] However, as discussed above, by August 22nd Duff was no longer in the area.

Second are the tombstones in a private cemetery where Jacob Turknette and John S. C. Turknette are buried. This epitaph claims the two men were killed in 1862. This is not correct: they were killed in the spring of 1863. The tombstone reads, "Beat to death by bullwhips by the men of Col. James Duff." [43] By the spring of 1863, Duff had been gone from the Hill County for months. These two Turknette men were killed by members of James M. Hunter's Company A, the Frontier Regiment, which generally were the same men of Davis' 1862 Company. [44] The main point here again; James Duff had been gone from the Hill Country for over six months.

Next, also in a private cemetery, the epitaph of Philip Brandon Turknette reads "Murdered by Col. Duff's men in 1862." [45] The problem here is Turknett was not killed until July 13, 1864, by one Jonas Harrison, a member of William Banta's Company A, Frontier Regiment – the successor of Hunter's company. Harrison was convicted of the murder and spent several years in the state penitentiary at Huntsville. [46] Again, this murder was committed years after Duff left the Hill Country.

Finally, in the Fredericksburg Cemetery is yet another headstone for the Itz brothers; Heinrich and Jakob. They were killed in February, 1863, by men of James Hunter's Company A, Frontier Regiment. On their tombstone the epitaph reads *"Von Der Duff Bande Ermordet* (killed by Duff's Band of Murders)."[47] As in the four cases above, these men were killed long after Duff left the Hill County by troopers of Company A, Frontier Regiment under three different commanders, Davis, Hunter, and Banta.

On all the tombstones listed above, the dates of death are wrong. They were killed long after Duff left the Hill Country, and all of the tombstones were erected by family members long after their deaths. These family members relied on family stories and folklore. Such is the reliability of family stories and folklore!

Now, let's return to Duff's 'Reign of Terror.' There are four elements of this reign of terror myth. First was James Duff's arrest of three insurgent leaders; Philip Braubach, Frederick Lochte, and F. W. Doebbler and Duff's search for Jacob Kuechler. Second, in 1861 and mid-1862 there was a "Reign of Terror." This was during the height of the league's influence. The individuals conducting this "Reign of Terror" were the Unionists and insurgents. They raided farms and businesses of Confederate supporters to include Robert Gibson's farm [48] and Thomas Doss's mill where shots were fired at one of his slaves. [49] They intimidated several other area citizens. They beat several pro-Confederate men, to include Oscar Basse, and Basse's friend, who tried to come to his aid. A third man was beaten and a fourth man "was pulled from his horse and broke his head." [50] The Unionist insurgents warned Joseph Poetsch not to align himself with the Confederates. [51] Philip Braubach, the insurgent sheriff, arrested Charles Schwarz because he enrolled in Captain Van der Stucken's Confederate company. [52] Two State company commanders, Charles de Montel and Henry Davis, sent reports that the insurgents in the mountains were claiming they were going to raid Frontier Regiment camps. [53] The insurgents let it be known, they were going to free Federal prisoners being held at Camp Verde. [54] They attempted to kill Captain Van der Stucken [55] and did kill a man named Basil Stewart for informing on them. [56] Fear among Gillespie citizens, both Unionist and Confederate alike, became so great that nine men prepared wills or deeds giving their property to friends or family members. [57] The area's southern sympathizers sent requests to General Bee to send troops for their protection. [58] General Bee responded to the State Company commanders and the area citizen. He sent two battalion task forces into the area. [59] Now the Unionists and insurgents began to feel they were the ones in danger and decided to flee and join the Union Army, resulting in the Nueces Battle on August 10, 1862, in which nineteen insurgents were killed. [60] Another seven were killed on October 18, 1862, in a second attempt to reach Mexico. [61] An eighth man died from his wounds inflicted on October 18th. [62] Now came the forth phase of the "reign of terror" when nine more Nueces Battle survivors were killed. [63] The deaths of all these thirty-six men have been placed on James Duff, but of those thirty-six, Duff may have had something to do with the deaths of just two, Bock and Tays.

There was another element of the 'Reign of Terror,' in the 1863 and 1864 time-frame; thirty-four men killed in what was called the 'Bushwhacker War' in the Hill Country. [64] So, during the Civil War, at least seventy men were killed in the Hill Country. This could well support the hypothesis that there was a 'Reign of Terror.' But James Duff had very little to do with either the 'Reign of Terror' or the Bushwhacker War; those men were not killed during his tenure in the Hill Country.

There are two documents Duff 'bashers' refer to in claiming Duff was responsible for the deaths of so many men. First is a Howard Henderson letter written on October 16, 1908, and second is a handwritten note by John Sansom listing individuals Duff killed. Henderson's letter stated, "I am now on my death-bed. I know that J. M. Duff and his company of murderers killed many of my neighbors and friends. My uncle and cousins Schram Henderson, my wife's father and brother,

Turknette, were murdered; my neighbors, Hiram Nelson, Frank Scott and his father, Parson Johnson and old man Scott were all butchered by Duff and his gang." [65] This letter was written by someone who was in the Nueces Battle and knew first-hand the names of men executed. However, it is full of errors. If these men were even executed, it was done long after James Duff left the Hill Country.

Let us examined this letter; Henderson's uncle was Sebird Henderson, executed on or about August 23, 1862, at Spring Creek. Two others whom Howard Henderson identified correctly were executed about August 23, 1862; Hiram Nelson and Frank Scott. The names of the 'cousins' are unknown, and it is not known who Schram Henderson was. No such man has been identified as a member of the Henderson family. Frank Scott's father was Benjamin Scout who was not executed. It is also uncertain who 'old man Scott' was, although one John Scott, then the Burnet County chief justice, was executed about the time the Spring Creek murderers were committed. Parson Johnson can not be identified, but there are no records of any by that name being executed. Henderson's wife was Narcissa Turknette. Her father was Jacob Turknette and her brother was John Turknette. Both were killed in the spring of 1863, as discussed above.

The Sansom handwritten note identifies nine men whom Duff's regiment "rode over the country . . . (and) put to death during the last days of July (1862)." The nine names include Gustav Tegener, Young Turknette, Rev. Tom Scott, Frank Scott, Rev. Jim Johnson, Hiram Nelson, Warren Cass, Wm. Schultz, (and) Ephraim Henderson. [66] Gustav Tegener, Frank Scott, and Hiram Nelson were three of those four men whom Davis or Donelson hanged on August 23, 1862, at Spring Creek. Young Turknette was executed in July of 1863.

Warren Cass was hanged on March 4, 1864, by Captain William Banta. William Schultz is a mystery. A William Schultz was living in Comfort or Sisterdale in 1862 and may have been a member of the Union Loyal League. A William Schultz enrolled in Duff's Company E on November 1, 1862. It is believed he deserted and on August 5, 1863, enrolled in Company D, First Regiment (Union) Texas Cavalry. Schultz deserted from Company D on May 17, 1864. It is also possible Sansom is referring to Louis Schuetze who was hanged on February 1, 1864, just north of Fredericksburg. Again, it is not known who Schram Henderson was.

ENDNOTES – Myth # 11

1. Handout, Comfort Heritage Foundation, Inc, n. d. ca 1995.

2. General Order Number 45, Headquarters, Department of Texas, Houston, Texas, May 30, 1862, 'War of the Rebellion: A Compilation of the Official Records of the Union and Confederate Armies' (OR), Series I, Volume 9, pp. 715-716.

3. Report Brigadier General H. P. Bee, Headquarter Sub-Military District of the Rio Grande, San Antonio, Texas October 21, 1862, to Headquarters First District of Texas, San Antonio, Texas OR, Series I, Volume LIII, pp. 454-455.

4. Duff's Report, p. 786.

5. "Minutes of Meeting, Gillespie County Rifles, February 23, 1862, and March 29, 1862, with copy of the Gillespie County Rifles Resolution, District Clerk's Office, Fredericksburg, Texas.

6. Duff's Report, p. 786.

7. Ibid.

8. Biesele, Rudolph L., 'The Texas State Convention of Germans in 1854', *Southwestern Historical Quarterly*, Volume XXXIII No 4, April 1930, p. 251.

9. Records, 31 at Brigade District, A.G.C. TSA

10. Letter Julius Schlickum, dated December 21, 1862, to his father-in-law.

11. Duff's Report, p. 786.

12. *San Antonio Herald*, June 28, 1862.

13. Analysis of Duff's Report.

14. "Records of the Confederate Military Commission in San Antonio, July 2-October 10, 1862, Edited by Alwyn Barr, *Southwestern Historical Quarterly*, Volume LXXI, No. 2, October 1967 pp. 253-272.

15. Camp Pedernales Post Return for the Month of August, 1862, National Archives, Washington, D. C.

16. Bee's Report, p. 454 and Camp Pedernales Post Return for the Month of August 1862, National Archives, Washington, D.C.

17. Bee's Report, p. 454, and Camp Pedernales Post Return for the Month of August 1862, National Archives, Washington, D. C.

18. Bee's Report, p. 454.

19. www.militaryfactory.com/dictionary/military-terms.

20. Bee's Report, p. 454.

21. Ibid.

22. *Webster's New Collegiate Dictionary*, A Merriam-Webster G. & C. Merriam Company, Springfield, Massachusetts 1979, p. 1158.

23. Betzer, Roy J., *Early Fredericksburg and Fort Martin Scott*, (unpublished MS, n. d., copy in possession of author), p. 4.

24. Camp Pedernales Post Return for the Month of August 1862, National Archives, Washington, D. C.

25. Ibid.

26. Ibid.

27. Ibid.

28. Handout, Comfort Heritage Foundation, Inc, n. d. ca 1995.

29. Camp Pedernales Post Return for the Month of August 1862, National Archives, Washington, D. C

30. Ransleben, Guido E., *A Hundred Years of Comfort in Texas; A Centennial History*, (The Naylor Company, San Antonio, Texas, 1954), p. 119.

31. Harper Centennial Committee, *Here's Harper*, (Radio Post, Inc., Fredericksburg, Texas, 1980), p. 12.

32. Hopkins' Diary.

33. Ibid.

34. Hopkins' Diary;

35. Harper Texas Sesquicentennial Committee, *Here's Harper Two*, (Nortex Press, Austin, Texas, 1986), p. 327

36. J. W. Seal's Letter of September 8, 1862 and Henry Schwethelm's Letter of 1913.

37. Letter, Captain James Duff, Commanding Company of Texas Dragons to Captain C. John Mason Acting Assistant Adjutant General, Depart of Texas, August 25, 1862. Copy located in James Duff's Confederate Service records; and Letter John Donelson, Provost Marshal, Camp Pedernales, August 21, 1862.

38. Kendall County District Court Records, Case Number 5.

39. Duff's Letter of August 25, 1862.

40. Treue Der Union Monument, Comfort, Texas, and Siemering, *Germans During Civil War*.

41. Ibid.

42. Harper Sesquicentennial Committee, *Here's Harper Two*, p. 327.

43. Ibid., p. 328.

44. The men who killed the Turknettes were: William Paul, John Paul, James Glenn, and Joseph Glenn. Statement by John Larremore, April 2, 1864, Gillespie County District Records.

45. Harper Sesquicentennial Committee, *Here's Harper Two*, .p. 327.

46. Gillespie County District Records, Case Number 5.

47. Heinrich and Jakob Itz tombstone, Der Friedhof Cemetery, Fredericksburg, Texas, Section 18, Graves 12 and 13.

48. Letter, R. A. Gibson, Camp Davis, March 31, 1864, A.G.C., TSA, Austin, Texas; Duff's Report pp. 785-786.

49. Gillespie County District Court Records, Case 101.

50. "Records of the Confederate Military Commission in San Antonio July 2 - October 10, 1862" (CMC). Barr, Alwyn (Ed), *Southwestern Historical Quarterly*, Volumes LXX (July 1966), pp. 93-109; LXX (October 1966), pp. 289-313; LXX (April 1967), pp. 623-644; LXXI (October 1967), pp. 258-263, 267-268 & 272-277; LXXIII (July 1969), pp. 83-90; and LXXIII, pp. 243-274.

51. Ibid.

52. Ibid.

53. Letter, Captain Charles de Montel, Camp Verde, August 3, 1862 to Colonel James Norris, A.G.C., TSA Austin, Texas and Letter, Captain Henry Davis, July 25, 1862, to Colonel James Norris, A.G.C., TSA, Austin, Texas.

54. Bitton, Davis (Ed.), *Reminiscences and Civil War Letters of Levi Lamoni Wright*, (University of Utah Press, Salt Lake City, Utah), 1970, pp. 23-24, 1970, and Siemering, *Germans During Civil War*, June 1, 1923.

55. Ibid.

56. Weber, Adolf Paul. *Die Deutsche Pioniers Zur Geschichtes des Deutschthums in Texas*, San Antonio, Texas: Published by Adolf Paul Weber, 1894 and Siemering's *Germans During Civil War*, June 1, 1923.

57. These nine included: Friedrich Wilke, Casper Fritz; Henry Heimann, Heinrich Markwordt, Jacob Dearing, Edward Felsing, Ernst Cramer, Ferdinand Simon, and John W. Sansom. Deed Records of Gillespie County Texas, Volume H, pp. 183-187; Deed Records of Kendall County, Texas, Volume I, pp. 46-53; and Kerr County Probate Records, Volume B, pp. 58-59.

58. Report Brigadier General H. P. Bee, Headquarters Sub-Military District of the Rio Grande, San Antonio, Texas October 21, 1862, to Headquarters First District of Texas, San Antonio, Texas OR, Series I, Volume LIII, pp. 454-455. Also see Siemering, *Germans During Civil War*, June 1, 1923.

59. Ibid. and Lieutenant Colonel Nat Benton's Report, *San Antonio Herald*, August 30, 1862.

60. These nineteen included Leopold Bauer, Fritz Behrens, Ernst Beseler, Louis Boerner, Albert Bruns, Hilmar Degener, Hugo Degener, Pablo Diaz, John H. Kallenberg, Heinrich Markwordt, Christian Schaefer, Louis Schieholz, Emil Shreiner, Heinrich Steves, Wilhelm Telgmann, Michael Weyrich, Heinrich Weyershausen, Adolph Vater, and Fritz Vater.

61. These seven included Peter Bonnet, Joseph Elstner, Edward Felsing, Henry Hermann, Valentine Hohmann, Franz Weiss, and Moritz Weiss.

62. This eighth was Fritz Lange, whose name is not on the *True Der Union* monument, which means he died after 1866.

63. These nine included Wilhelm Boerner, Conrad Bock, Theodore Brunkish, Herman Flick, August Luckenbach, Adolph Ruebsamen, August Ruebsamen, Heinrich Stieler, and Fritz Tays.

64. These thirty-four men included, James Billings, John Blank, Peter Burg, Michael Burtcher, Warren Cass, Amos Fairchild, William Feller, Joseph Fries, Henry Hartman; _____ Hines, Henry Itz, Jacob Itz, Henry Kirchner, Johann Klein, Mr. Lundy, William Lundy; Xavier Neel, J. Monroe Nixon; Johann Pletz; Coston J. Sawyer; William Martin Sawyer; John Smart, Moses Moran Snow, William F. Snow, Jesse Starr, George Thayer, Jacob Turknette, John Turknette, Philip B. Turknette, _____ Van Winkle, Benjamin Watson, Zack Whittington, _____ Williams, and Columbus P. Wood.

65. Letter. Howard Henderson, October 1908, Ingram, Texas contained in Ransleben's *Hundred Years of Comfort*, pp. 119-120.

66. John Sansom's Ledger, located in Sansom's File, DRT Library at the Alamo, San Antonio, Texas and Sansom, *German Citizens,* also located at the DRT Library at the Alamo, San Antonio, Texas.

12 – Duff Threatened to Hang All 'Dutchmen'?

MYTH: While he was provost marshal Duff openly stated, "The God damn Dutchmen are Unionists to a man," and "I will hang all I suspect of being anti-Confederate." [1]

FACT: There is no documented evidence Duff ever made such a statement.

DISCUSSION: The first time this myth was recorded is in T. R. Fehrenbach's book *Lone Star*, published in 1968. Since then, the quote has been used countless times, but the source has always been Fehrenbach. During the 'Nueces Encounter' Symposium on March 22, 1997 Mr. Fehrenbach was asked for his source for the quote. In a letter to the author dated March 26, 1997, he was unable to provide a source. [2] Based on several likely sources Mr. Fehrenbach stated that it likely came from a German language newspaper, either the *New Braunfelser Zeitung* or the *San Antonio Freie Presse fuer Texas* or an oral interview. Despite an intensive search of these papers, this quote has not been located. It is the author's opinion this quote, like many others, is incorrect and if does exist was printed in a biased article without a source. The only known primary source expressing sentiments anywhere near this quote is contained in R. H. Williams book *Ruffians*: "Fredericksburg was a town of about 800 inhabitants, almost all of them Germans and Unionists to a man." [3]

It should be remembered Duff was the provost-marshal stationed at Fort Martin Scott. He had no troops under his command. All of the hangings took place near Spring Creek and the camps of Captain Davis and Captain Donelson.

Wm. Paul Burrier

ENDNOTES - MYTH #12

1. Fehrenbach, *Lone Star*, p. 363.

2. Letter, T. R. Fehrenbach to author, San Antonio, Texas March 27, 1997.

3. Williams, *With the Border Ruffians*, p. 232.

13 – Unfit For Command?

MYTH: James Duff was totally unfit for command; while an enlisted member of the U. S. Army he committed a grave crime. As punishment he was court-martialed, found guilty, tied to a whipping post, whipped, and drummed out of the army in disgrace. He kept his crime a secret and was able to get a Confederate commission only because of his political influence. [1]

FACT: It is true that Private James Duff, while a member of the regular army, was court-martialed for the modern equivalent of absent without leave (AWOL). On the night of May 14/15, 1849, Duff and four other enlisted troops left their station, Fort Gibson, without authority. They were apprehended on May 22, 1849, and returned to Fort Smith. All five were court-martialed, found guilty, and sentenced to "reimburse the United States the expenses of $30.00 for their apprehension; forfeit all pay and allowances due them, except just debts to the Sutler and Laundress; be confined at hard labor in charge of the guard, with a chain two to three feet long, attached to each leg with a shackle, for a period of six months; at the expiration of the six months to receive fifty lashes on his bare back laid on with raw hide; be indelibly marked on the left hip with the letter D one and a half inches long; have his head shaved; and be drummed out of the service." [2]

A note of explanation is needed at this point. At this time army punishment was severe. But the offender could be pardoned if he proved to be good soldier, which was the case of Duff and three others.

DISCUSSION: The major source on Duff's court-martial is an account by General Davis S. Stanley. He was the Federal commander of troops in Texas in the 1880s and supported the individuals who had been Unionists during the war. As such he knew Eduard Degener, August Siemering, Ernest Cramer, Fritz Tegener, Philip Braubach, and Jacob Kuechler, who became prominent politicians after the war and during Reconstruction. John Sansom even goes to the extreme of saying Stanley told him he witnessed the entire affair to include the whipping, branding, and being drummed out of service. Stanley 'may have' made statements such as these to be accepted in the Texas political arena. Many historians and writers rely on the Stanley account to prove Duff's bad character. The problem with such a statement is that it is not true. Stanley never made such a statement! General Stanley does mention the incident in his memoirs, but nearly everyone omits Stanley's last sentence which is; "but he was pardoned by General Arbuckle." [3]

James Duff, as well as three of the four soldiers, had their sentences remitted. On October 29, 1849, General Matthew Arbuckle, the Commander of the 7th Military Department, issued Special Order Number 9, which remitted the remaining portion of their sentences. The men were required to pay the expense of their apprehension. The three soldiers returned to duty. They did not receive fifty lashes on their backs, were not marked with a 'D,' and not drummed out of the service. Duff remained in the army and had an outstanding career. [4]

On November 14, 1849, Duff was transferred to Company K, 5th U. S. Infantry Regiment, commanded by Brevet Major Nathan Beakes Rossell. Less than three weeks later, on December 1, 1849, Duff was promoted to Third Corporal. On June 1, 1850, Duff was promoted to Fourth Sergeant and appointed the acting Quartermaster Sergeant. [5] James Duff received his honorable discharge on January 6, 1854. [6]

ENDNOTES - MYTH #13

1. Sources stating this as a fact are: Sansom, *Battle of Nueces*, p. 2; Fehrenbach, *Lone Star*, p. 363; Ransleben, *A Hundred Years of Comfort*, p. 105; Biggers, *German Pioneers*, p. 59; Wooster, Ralph A., *Texas and Texans in the Civil War*, p. 114; Glenn, Frankie Davis, *Frontier Series: John W. Sansom*, John W., *Battle of the Nueces*, 1991, pp. 5-6; McGowen, *Horse Sweat and Powder Smoke*, p. 66; Smith, *Frontier Defense in the Civil War*, p. 156; Shook, *The Battle of the Nueces*, p. 32; Rutherford, *Defying The State of Texas*, p. 17; Clare, *Bloody Ground*, p. 49; Kelton, Texas Germans Fell on Nueces; Alberthal, *Bushwhackers In Them Thar Hills*; Schmidt, *Publication Commemorating The 50th Anniversary Of The Battle On The Nueces*; Gold, *Gillespie County In The Civil War*, p. 27; Felgar, *Texas in the War for Southern Independence*, p. 342; Curtis, *History of Gillespie County*, p. 57; Hall, *Texas Germans in State & National Politics*, p. 87; Weinheimer, *Fredericksburg During Civil War*, p. 56; Heintzen, *Fredericksburg During Civil War*, pp. 43-44; Comfort Heritage Foundation Handout; Baulch's The Dogs of War Unleashed, p. 130; and Dykes-Hoffmann, *German Texan Women on the Civil War Homefront*, p. 68.

2. Orders Number 10, Headquarters 7th Military Department, Fort Smith, June 9, 1849; Orders Number

3. Stanley, D. S., *Personal Memories of Major General D. S. Stanley*, Harvard Press, Cambridge, Massachusetts c. 1917, p. 234.

4. General Order Number 9, Headquarters 7th Military Department, Fort Smith, October 23, 1849, Records of the Adjutant General's Office (Record Group 94), National Archives, Washington, D. C.

5. Muster Rolls Company K, 5th U. S. Infantry, December 31, 1849 – February 28, 1850; February 28, 1850 – April 13, 1850; April 13, 1850 – June 13, 1850; June 13, 1850 – August 31, 1850; August 31,1850– October 31, 1850; October 31, 1850 – December 31, 1850; December 31, 1850 – February 28, 1851; February 28, 1851 – April 13, 1851; and April 13, 1851 – June 13, 1851, Records of the Adjutant General's Office (Record Group 94), National Archives, Washington, D. C.

6. Muster Rolls Company K, 5th U. S. Infantry from Period August 31, 1853, to February 28, 1854. Records of the   Adjutant General Office (Record Group 94), National Archives, Washington, D.C.

14 – Did the Loyal League Really Stand Down?

MYTH: With the arrival of "regular' Confederate troops in the Hill County, the Union Loyal League disbanded the three companies and issued an assurance that no armed conflict was to be expected. [1]

FACT: The League did not disband its' three companies and no assurance was given that no armed conflict was to be expected. [2] At the time this alleged statement was made to Sansom there were no Confederate troops in the area to give assurance that no armed conflict was to be expected.

DISCUSSION: The only time this statement is made is in Sansom's account. Nevertheless almost every account written about the Nueces Event includes this statement. It is always Sansom's statement that is quoted as the source. The State and Confederate authorities did not know of the existence of the companies. No insurgent account makes this statement. This is just another example by 'historians' and 'writers' to demonize the secessionists.

Several leaders of the League's military wrote accounts of the events. None of them say anything about disbanding their companies or an assurance that no armed conflict was to be expected. Such accounts include those of Fritz Tegener, Henry Schwethelm, Ernst Cramer, August Siemering, Julius Schlickum, August Hoffmann, Fritz Schellhase's journal, Schutze's article, "Was a Survivor of Nueces Battle," Helen Raley's newspaper account "Blackest Crime in Texas Warfare" Eduard Degener's letter, and the Usener brothers' story. [3]

Wm. Paul Burrier

ENDNOTES - MYTH # 14

1. Sansom, *Battle of Nueces*, p. 3.

2. There is no evidence, except in Sansom's book that the companies were to disband and an assurance was given that no armed conflict was expected.

3. Letter, Fritz Tegener, Austin, Texas, to Herr August Duecker, Gillespie County, Texas August 23, 1875; Letter, Henry Schwethelm, May 16, 1913 to his grandson, Otto Schwethelm; Letter, August Hoffmann to his children, September 1, 1925; Letter, Ernst Cramer to his parents, October 30, 1862; Siemering, August, *Des Lateimische, in Texas,* 1874; Siemering, A., *The Germans in Texas during the Civil War*, contained in San Antonio *Freir Presse fuer Texas*, May-June 1923; August Siemering, *Ein Verstehtes Lebdn,* 1876; Fritz Schellhase Journal, c1912; Albert Schutze-Was a Survivor of the Nueces Battle, 1924; Helen Raley, Blackest Crime in Texas Warfare, in *Dallas Morning News* May 5, 1929; Letter, Eduard Degener, August 1, 1862; Usener, Raymond, *Jacob and Ludwig Usener Story*, 1997; and F. W. Schweppe's article, Bonnet Brothers, not dated.

15 – The Confederate Amnesty?

MYTH: The Texas Governor, or some other State or Confederate authority, issued a proclamation about the middle of July 1862, stating anyone who did not want to live under Confederate rule had 30 days to leave the state. The August 1862 fleeing insurgents were traveling under the authority of this proclamation. [1]

FACT: No Confederate national or state authority issued any such proclamation in the summer of 1862, permitting anyone who did not want to live under Confederate rule 30 days to leave. Texas Governor Edward Clark <u>had</u> issued a proclamation on June 8, 1861, (a year before later writers say that one was issued) telling everyone that they had twenty days to take the Confederate Oath of Allegiance or leave the state. [2] Confederate President Jefferson Davis issued a proclamation on August 14, 1861, (again, a year before 'historians' and 'writers' claim there was one issued), ordering everyone living in the territory of the Confederate States to either take the Confederate Oath of Allegiance within forty days or leave; otherwise they would be treated as 'alien enemies.' [3]

DISCUSSION: This myth is so deeply imbedded in the Nueces Battle/Massacre Story, that every modern historian and writer revisiting the matter states this as a fact. Almost all rely on Sansom's 1905 pamphlet or Biggers' 1925 book *German Pioneers in Texas* as sources. (Endnote # 1 of this chapter has a detailed list of examples of the stating of this myth.) One account goes so far as to claim the insurgents still had 12 days remaining. [4]

In 1994 I began researching the background of the Nueces Battle and massacre of August 10, 1862. At that time I had no reason to believe that what had been written was incorrect. I grew up in the Texas Hill Country and had heard about the event all my life. One of the criteria I established when I stated researching was to go to the primary source for information rather than relying on what someone else had said or written. When I arrived at the part of the story of the fleeing insurgents and began reading the stories about this proclamation, this became an item to look up at the state archives. It soon became apparent there was no record of any such proclamation: not in the Governor's Proclamation Book or in any file in the archives of any Texas or Confederate records. I checked in Official Records of the Civil War (OR), but no record. I checked the newspapers of the time, both English and German, but no record. By now I realized there was a problem. I rechecked the accounts. In every case the source that was quoted was Sansom's 1905 pamphlet: "The Battle of Nueces in Kendall County." These were what I considered primary accounts; surely they would have a source? It appeared that when the author wrote about the proclamation, they used either Sansom's account or what the writer believed; and so it would be included in subsequent interviews and recollections. Such accounts included Sansom's 1905 pamphlet; a 1923 Albert Schutze interview with Henry Schwethelm, and a 1929 interview by Helen Raley of the *Dallas Morning News* with August Hoffmann. It was about this time I began to realize almost everything that had been written about the Nueces Event of August 10, 1862, was incorrect, very biased, and full of myths, not facts.

Most of the other myths could be addressed by finding actual documents giving the correct data. However in the case of the alleged proclamation I was going to have to prove a negative, to wit: "No proclamation was issued by any State or Confederate authority giving the insurgents 30 days to leave the state sometime in the summer of 1862."

I was aware of the two proclamations issued in mid-1861 which basically said what writers claimed of a proclamation issued in the summer of 1862. These two proclamations were the Texas Governor Proclamation of June 8, 1861, telling everyone they had twenty days to take the Confederate Oath of Allegiance or leave the state; and Confederate President Davis' Proclamation of August 14, 1861, requiring any male in the Confederacy to take the oath of allegiance within forty days or leave the South.

For those critics who claim, "Well maybe the settlers were not aware of these two proclamations," [5] the period Texas newspapers, both English and German, intensively reported these proclamations. For example, President Davis' Proclamation was published in English on August 29, 1861, in the *San Antonio Daily Ledger and Texan,* [6] the *San Antonio Weekly Ledger and Texan* on August 31, 1861, [7] the *Austin Semi-Weekly News* on August 29, 1861, [8] and the *Austin State Gazette* on August 31, 1861, [9] and in German in the *Neu Braunfelser Zeitung* on September 6, 1861. [10] Hill Country citizens read these newspapers and were well aware of Davis' Proclamation, and many chose to take the oath. For example; in the Comfort area, thirty-one did so. [11]

Could it be the Nueces Battle survivors were claiming they fled under Davis' Proclamation of August 1861? Surely not! Davis' Proclamation had expired almost a year before! Would they have the gall to make such an outrageous claim? If they did, then what else in their accounts was suspect?

I began digging into what I had thought were primary accounts and try to pinpoint when these claims begun to surface. One account I looked at, was Robert Williams' book *With the Border Ruffians.* [12] He was extremely critical of everything Duff did. If the insurgents were travelling under any type of authority such as the proclamation, he would certainly have said so, yet there is no mention in his book about such a proclamation. Other accounts of the battle and events surfaced. Several were written in German and shortly after the battle.

The first was Ernest Cramer's letter of October 30, 1862, to his parents. [13] Cramer had been the captain of the Kendall Company of the league's military battalion. Cramer made no reference to any proclamation. The second was the Julius Schlickum letter of December 21, 1862, to his father. [14] Schlickum made no reference to any proclamation. The third was Fritz Schellhase's journal. [15] It is not dated but written shortly after the August fleeing group left for Mexico. Schellhase made no reference to any proclamation. The next source was an account of the Bonnet brothers who were in the October 1862 fleeing insurgent party. [16] There was no reference to any proclamation. The sixth was a long newspaper article written by August Siemering in 1875. [17] He was an original member of the League and made no reference to any proclamation. The seventh was the August 23, 1875, letter of Fritz Tegener, who was the commander of the August fleeing group. [18] Like the others, Tegener made no reference to any proclamation.

From these letters written shortly after the battle, it was clear the earliest accounts did not claim the insurgents were travelling under the authority of any proclamation. When did this myth begin to appear? After detailed searching, I found the first mention in print was at the twenty-fifty anniversary of the battle, in an August, 1887, Jacob Kuechler letter that was read at the anniversary observances. In it Kuechler said, "Under the proclamation we had plenty (of) time to leave Texas." [19] In an 1894 German-language book quoting Kuechler there is a different version. Kuechler says in the 1894 book, "It is not to be forgotten that the Confederation at the outbreak of the war granted each citizen who did not want to stay under the Confederate flag, 90 days to leave." [20] It can be concluded that Kuechler was less than truthful in his August, 1887, letter and the proclamation referred to was the President Davis Proclamation of August, 1861.

The next creditable reference to a proclamation the insurgents were travelling under, is John Sansom's 1905 account of the Nueces Battle. In this account Sansom says, "Having read a proclamation from the Confederate Government announcing that all persons not friendly to it might leave the county, we believed we had a right to go." [21] In a 1911 article Sansom again addressed a proclamation. In it he spoke about many Unionists reaching the south side of the Rio Grande: "Called the attention of the Confederate military authorities to the fact that there were many other self-Unionists; which it might be well to watch and punish for the their failure to leave the State, as by proclamation already issued all persons unwilling to take the oath of allegiance to the Confederacy had been commanded to do." Sansom goes on to tell about why so few obeyed the proclamation. Some of the reasons were, "The time allowed for their departure was short; nobody cared to pay a fair price for land and other property that when forfeited to the Confederacy would be auctioned off and sold for a song." Sansom further explained,

"Without money in Mexico, the nearest place to which he must go, what could a refugee hope for? . . . If he left, what hope had he that it would escape the scalping knife?" He further explained he was but one against many and had the "attractions of a home blessed by a charming young wife and the work I had to do as a farmer, stockraiser and occasional surveyor kept me home." He felt that, "All things considered he resolved to stay, not to abandon his property, not to desert his family and leave it to the prey of the savage . . . Certainly" Sansom admitted, "if he stayed quietly at home attending to his business and doing nothing against the Confederacy, he would not be molested. Arguing thus, very few of those to whom the proclamation applied obeyed its commands." [22] Sansom said this worked well until June, 1862. So it can be concluded that the proclamation Sansom was referring in his 1905 book is President Davis' proclamation of August, 1861.

Another Nueces Battle survivor was Henry Schwethelm, whom Albert Schutze interviewed in August, 1924. The Schutze interview has Schwethelm saying, "The Governor of the State issued a proclamation that all persons who would not take the oath of allegiance to the Confederate States would have to leave the State within thirty days." [23] The only other Schwethelm document located about the proclamation was a May 16, 1913, letter to his grandson, Otto. In that letter Schwethelm says, "The governor of the State issued a Proclamation that all that would not take the oath of allegiance of the Confederacy had to leave the State within 30 days, so we left." [24] Of all the articles written by or about an insurgent, Schwethelm is the only one to repeat the claim that a proclamation was issued in 1862 in more than one document.

The final Nueces Battle survivor who appears to say the August group was travelling under a proclamation, giving them time to leave Texas is August Hoffmann. Like the

Schwethelm account, Hoffmann's account is contained in a newspaper interview. This interview was by Helen Raley which appeared in the *Dallas Morning News* on May 5, 1929. In the account Raley quotes Hoffmann saying, "Trusting in this (proclamation) many went (back), although the time set in the Governor's proclamation had expired ten days before." [25] Hoffmann was referring to the fact that several insurgents who had gathered at Bushwhack Creek, left the group before the main body started for Mexico. Since the fleeing group left Bushwhack Creek about August 3rd, this means Hoffmann was saying the proclamation had expired about July 24, 1862, which means the proclamation was issued about June 24, 1862. There are several items in the Raley interview where she seems to be quoting Hoffmann. Other accounts show Hoffmann never made many of these statements. There is a second Hoffmann account, written by Hoffmann himself on November 1, 1925, or four years before the Raley interview. [26] This letter is very detailed about events besides the Nueces Battle. Nowhere in this letter does Hoffmann say anything about a proclamation. This casts serious doubt that Hoffmann ever made the statement contained in Raley's account.

What all this comes down to is that – except for Schwethelm account – no survivor of the Nueces Battle claims that they were travelling under a proclamation. Therefore, it can be stated with confidence that no such proclamation was issued by either the State or Confederate authorities in mid-1862!

ENDNOTE - THE MYTH #15

1. Letter, Jacob Kuechler, to James Newcomb, August 1887, read at the Twenty-Fifth Anniversary of Nueces Battle; Weber, Adolf Paul, *Deutsche Pioniere, Zur Geschichte des Deutschthums in Texas*, (Selbstverlag des Verfassers Press, San Antonio, Texas, 1894). Sansom, John W., *Battle of Nueces River, August 10, 1862*, (Privately published, San Antonio, Texas, October 1, 1905); Schmidt, Eduard, *Commemorating The 50th Anniversary Of The Battle On The Nueces, August 10, 1862*, (Kerrville, Texas, August 1902); Letter, Henry Schwethelm to his Grandson Otto, May 1913; Biggers, Don H., *German Pioneers in Texas*, (facsimile reproduction by Fredericksburg Publishing Company, Fredericksburg, Texas, 1925), p. 58; Albert Schutze Was a Survivor of the Nueces Battle, *San Antonio Express*, August 31, 1924; Helen Raley, The Blackest Crime in Texas Warfare, *Dallas Morning News*, May 29, 1929; Diamond Jubilee Souvenir Book of Comfort, Texas Commemorating 75th Anniversary August 18, 1929 of The Battle of Nueces by unknown author. By the late 1920s, the myth that a proclamation had been issued giving anyone who did not want to remain under Confederate rule had 30 days to leave Texas began appearing in Master and Ph.D papers. Some of these include: Felger, Robert Pattison, *Texas in the War for Southern Independence 1861-1865*, (University of Texas, Doctor of Philosophy, Austin, Texas, 1935); Hall, Ada Maria, *The Texas Germans In State and National Politics, 1850-1865*, (M. A. Thesis, University of Texas, Austin, Texas, 1938); Curtis, Sara Kay, *A History of Gillespie County, Texas, 1846-1900*, (M. A. Thesis, University of Texas, Austin, Texas, 1943); Frank W. Heintzen, *Fredericksburg, Texas, During The Civil War and Reconstruction*, (M.A. Thesis, St. Mary's University of San Antonio, 1944); Weinheimer, Ophelia Nielsen, *The Early History of Gillespie County, Texas*, (M. A. Thesis, Southwest Texas State Teachers College, San Marcos, Texas, 1952).

2. Texas Governor's Proclamation Book, TSA, Austin, Texas and San Antonio Herald, June 8, 1861.

3. 'An Act Respecting Alien Enemies', approved August 8, 1861, OR, Series II, Volume II, pp. 1368-1369 and 'An Act to Alter and Amend an Act Entitled "An Act For The Sequestration of the Estates, Property, and Effects of Alien Enemies and for Indemnity of Citizens

of the Confederate States, and Persons Aiding the Same in the War With The United States', August 13, 1861, OR, Series II, Volume II, pp. 932-944.

4. Raley, Blackest Crime in Texas Warfare. There is serious doubt that Hoffmann made this statement. It is more likely that Raley added it in the article, as she did several other items.

5. Thorpe, Helen, "Historical Friction", *Texas Monthly*, Austin, Texas, October 1997, p. 78.

6. *San Antonio Daily Ledger and Texan*, August 29, 1861.

7. *San Antonio Weekly Ledger and Texan*, August 31, 1861.

8. *Austin Semi-Weekly News*, August 29, 1861.

9. *Austin State Gazette*, August 31, 1861.

10. *Neu Braunfelser Zeitung*, September 6, 1861.

11. Kerr County Probate Book A, pp. 46-48, 50-54 and Kerr County Naturalization Records, pp. 92-109. This record was kindly provided to the author by Gregory Krauter of Comfort, who at a great expense located and obtained the ledger.

12. Williams, R., *With the Border Ruffians*.

13. Ernest Cramer's Letter.

14. Letter ,Julius Schlickum, dated December 21, 1862, to his father-in-law.

15. Schellhase Journal.

16. Boerner, Jur Bernhard, *The Bonnet Family From Chambons In The Dauphine*, (An English Translation) in 'Deutfsches Geschlechterburh', Volume 60, 1, 1928, copy provided by Esther Strange of Kerrville, Texas, a Bonnet descendent; Kerr County.

17. Siemering, *Germans in Texas During the Civil War, Freir Presse fur Texas*, May 4, 1923; May 8, 1923; May 11, 1923; May 15, 1923, May 22, 1923; May 25, 1923; May 29, 1923; June 1, 1923; June 12, 1923; and June 5, 1923.

18. Letter, Fritz Tegener, Austin, Texas to Herr August Duecker, Gillespie County, Texas, August 23, 1875, English   translated copy provided to author by Gregory Krauter.

19. Kuechler's Letter

20. Weber, *Die Deutsche Pioniere*, p. 12.

21. Sansom, Battle *of Nueces*, p. 11.

22. Sansom, John W., The German Citizens Were Loyal To The Union, *Hunter's Magazine*, November 1911

23. Schutze, "Was a Survivor of Nueces Battle".

24. Schwethelm's Letter.

25. Raley, "Blackest Crime in Texas Warfare".

26. Hoffmann's Letter.

16 – Duff's Spies Among the Unionists?

MYTH: Duff had several spies, namely Basil Stewart and Charles Burgmann, who kept him informed about the insurgent's anti-Confederate activities. [1]

FACT: Neither Duff nor Donelson had any spies within the Organization.

DISCUSSION: The German revolutionaries of 1848 had a tendency to suspect others of being spies in the pay of the government. When they arrived in the United States and Texas, this distrust of their fellow refugees remained. "Many of the exiles in (the United States and Texas) were too ready on the slightest evidence to charge one of their colleagues with being a spy," wrote a German-American historian. [2] In Texas this tendency was carried into their organization of the insurgency and creation of the Organization. There was an enduring fear that the Confederates had planted spies among them. When someone was discovered who had given information to the Confederate authorities, and who was a member of the Organization, he was banned as a spy who had been in the employment of the Confederate government all along. The insurgents claimed both Steward and Burgmann were spies, which they were not.

The first item that must be discussed is the definition of the words 'spy' and 'informant' and 'betrayer'. Webster defines 'spy' (*spionieren*) as "one who keeps secret watch on a person or thing to obtain information. One who acts in a clandestine manner or on false pretenses to obtain information in the zone of operation of a belligerent with the intention of communicating it to the hostile party." [3] Informant (*anzeigen*) is defined as "one who supplies cultural or linguistic data in response to interrogation by an investigator." [4] Betrayer (*verraten*) is defined as "to disclose in violation or confidence." [5]

Analysis shows neither Steward nor Burgmann meets the technical definition of spy. The term 'spy' implies giving reports on the individuals or groups which the Confederates wanted information about. It is very likely that Steward was a member of the American company of the Organization. At the time of Steward's enrollment in the Organization there were no Confederate troops in the area. The first such to arrive in the area were Captain James Duff and his Partisan Rangers in late May, 1862. Neither Duff nor any other Confederate could have recruited Steward. It is not known if Burgmann was a member of the Organization, but he could not have been recruited to become a spy because – as in the case of Steward – there were no Confederate troops in the area to report to. There is no evidence Burgmann contacted Confederate troops, prior to the arrival of Donelson's task force on or about August 1st.

Steward does not meet the definition of an informant because he was never captured and therefore interrogated. Since he was killed on July 5, 1862, he had to have turned himself in prior to that date. Just when he turned himself can be narrowed a little. Duff tells us that he arrived in Fredericksburg on May 28, 1862, and besides declaring martial law, he gave the citizens six days to take the Confederate Oath of Allegiance. [6] It was likely after this that Steward came in and took the oath. R H. Williams, a member of Duff's company, and Ernest Cramer, commander of the Kendall Company of the Organization provide us information about what happen next. Williams says, "Most of them did (take the oath), though some cleared out and took to the mountains rather than perjure themselves." [7] Cramer says, "Then it was that I began to really know people. Excepting a very few, all took the oath, and also betrayed their (Organization) officers. All officers had to immediately flee for their lives." [8] Thus it is likely that Steward came in after May 28 and before July 5.

Charles Nimitz, owner of the Nimitz's Hotel and captain of the Gillespie County Rifles, is one who does meet the definition of an informant. James Duff tells us about Nimitz and others who informed on the Unionists. Duff says, "On my return to Fredericksburg I found beyond doubt that the few citizens of the place who were friendly to this Government did not posses moral courage enough to give information to the provost-marshal of the sayings and doings of those who were unfriendly . . . I determined to summon them to meet with (Lieutenant Sweet) and myself—They obeyed the order and made affidavits in regard to certain citizens of the county." [9] R. H. Williams tells of the Gillespie citizens who informed on the insurgents, saying, "These prisoners . . . had been informed against by a Dutch Tavern-keeper in Fredericksburg." [10]

It is possible that Charles Burgmann meets the definition of an informant, based on a statement by R. H. Williams. He tells us, "One of the prisoners, an old soldier, and a friend of (Duff's) had been released, and he was to act as our guide and betray his friend, if possible, into our hands." [11] This took place in early August, 1862, after Donelson's task force arrived in Gillespie County. So Burgmann may have been captured and informed on the insurgents. First Lieutenant Colin McRae, the leader of the Confederate pursuit force, identifies Charles Burgmann as a civilian guide. [12] Weber, in his book *Die Deutsche Pioniere* tells us what Jacob Kuechler said, "A few days after they had started their march, they met a German by the name of Bergmann (sic) who had fallen upon a store of provisions in an open field. Perhaps as a joke or out of hunger, they took the food away from him, for which Bergmann, (sic) out of anger, put the pursuers on the trail of his countrymen." [13]

Both Basil Steward and Charles Burgmann fall into the category of betrayer. The insurgents condemned both to death. August Siemering tells of Stewart's betrayal, saying, "Here the spy was in the midst of the meeting. His name was Steward and he was an Englishman by birth . . . This man had nothing more important to do than to go to a judge and to report all that he had seen and heard. Luckily, he knew few by name but he had said enough to cause widespread alarm." Siemering goes on to tell of Steward's execution. "The leaders of the Unionists held a meeting and it was decided, unanimously, that this spy must die . . . The question was 'who would accomplish this?' It fell to a young man from Kendall County. The opportunity to carry this out presented itself immediately. "Steward, with the help of a Negro, were driving cattle from Comfort to the farm of his friend Attrill. The road led up the hills through a narrow canyon. As Steward proceeded through this canyon, the fatal shot fell, ending his life immediately." [14]

The insurgents were unable to execute Burgmann, because he was now with the Confederate forces, first as a guide and later as a soldier. Burgmann was hit in three places in the battle on August 10, 1862. [15] On October 1, 1862, he enrolled in Duff's Company B and remained with the unit until the end of the war. [16] After the war Burgmann fled to

Mexico. [17] Ransleben, in *A Hundred Years of Comfort in Texas* claims Burgmann was killed in Mexico by a Seminole Indian Negro, and his body thrown into the Rio Grande. [18]

But like most things regarding the Nueces Battle and Massacre there are yet other family accounts that tell of who the 'Betrayer' was and how he was killed. The Ferdinand Schulze family tells of a story Mr. Schulze told his family on his deathbed. He related "how he drew the straw to murder 'the informant' after the Civil War was over." [19] So who was 'the informant' that he killed? It was not Basil Steward, as we know when and how he was killed. It 'may have been' Charles Burgmann, but if it was, then the Ransleben's account about a Seminole Indian Negro killing Burgmann and throwing his body into the Rio Grande is wrong. [20] Splittgerber descendants claim that twenty-six-year-old Prussian native Oscar Splittgerber was one of those volunteers who executed the insurgents after the battle. He was named in a July, 1997, off-the-record interview. [21] This account and others relate that after Dennis Kingston, a soldier at Fort McKavett, was killed, his widow swore that at least ten Germans would pay with their lives for his death. About October of 1860, she married Oscar Splittgerber, who had worked for the Butterfield Stage Line and was a tanner for the Confederacy in Menard and Mason counties. Splittgerber joined the League using the alias 'Bargerum' (sic). According to this account, the Confederates captured him and, in order to save his life, he informed on the Organization's departure. He did so, not only to save his life, but also to fulfill his wife's oath. [22] Another account says Oscar Splittgerber was "the Charles Bergman who, on order of a Confederate officer at the Battle of Nueces, executed the eleven Unionists," and Oscar was murdered in 1889 for his actions during the Civil War." [23]

An Oscar Splittgerber family story says after the Civil War was over, the Splittgerbers and others were harassed by renegade northern soldiers because of their connection with the Confederacy and the family fled to Mexico, returning five years later. [24] Oscar Splittgerber settled first in western Menard County and later in Reeves County in west Texas.

Oscar Splittgerber's death occurred (one account has it by means of a bullet hole in the back of his neck [25]) on November 14, 1889, in Reeves County, and closely matches Frederick Schulz(e)'s deathbed confession. It is a fact that Oscar Splittgerber was killed on November 14, 1889, under mysterious circumstances. A detail adding credibility to a claim that insurgents killed Splittgerber is that Jacob Kuechler – an insurgent leader – was surveying for the railroad in west Texas in October, 1883. Kuechler might very well have recognized Oscar Splittgerber as one of the volunteers who killed the wounded insurgents, reported that sighting back to Comfort, and remnants of the Organization may have dispatched one of their own to execute the 'brayer'. [26]

However, it is not believed Splittgerber used the alias 'Charles Burgmann' and under that name helped kill the wounded insurgents. It is very possible Splittgerber joined the League under an alias (but not that of Burgmann), and when captured by the Confederates chose to betray the insurgents.

But what is clear is that the Confederates did not have an active spy in the Organization.

ENDNOTES - MYTH # 16

1. Weber, *Die Deutsche Pioniere,* Biggers, *German Pioneers,* p. 59.

2. Bruncken, E., Tolzmann, Don Heinrich, *The German-American Forty-Eighters 1848-1998,* (Indiana University Press, Indianapolis, Indiana 1998), pp. 43-46.

3. *Webster's New Collegiate Dictionary,* A Merriam-Webster G. & C. Merriam Company, Springfield, Massachusetts, 1979, p. 1119.

4. Ibid., p. 857.

5. Ibid., p. 104

6. Duff's Report, p. 785.

7. Williams, *With The Border Ruffians,* p. 237.

8. Cramer's Letter.

9. Duff's Report, p. 786.

10. Williams, *With The Border Ruffians,* p. 237.

11. Ibid.

12. Letter, Michael E. Pilgrim, Textual Reference Division, National Archives, Washington, D. C., January 31, 1995, containing a copy of Lieutenant C. D. McRae's Casualty List, Record Group 109, E 22.

13. Weber, *Die Deutsche Pioniere,* p. 13.

14. Siemering, *Germans in the Civil War in Texas,* June 5, 1923.

15. McRae's Casually List.

16. Charles Burgmann's Confederate Service Records, Record Group 109, National Archives, Washington, D. C.

17. Nunn, W. C., *Escape From Reconstruction*, (Texas Christian University, Fort Worth, Texas, 1956), p. 129.

18. Ransleben, Guido E., *A Hundred Years of Comfort in Texas*, p. 121.

19. Letter, Catherine Carrigan to Gregory Krauter, Comfort, Texas, August 1, 1991, copy kindly provided author by Gregory Krauter

20. Ransleben, Guido E. *A Hundred Years of Comfort in Texas*, p. 121.

21. Off the record interview, July 19, 1997 at Kerrville, Texas.

22. Mary Lewis Turner's oral presentation to the German-Texas Heritage Society at its September 5 -7, 1997 Convention at Kerrville, Texas.

23. Off the record interview, July 19, 1997, at Kerrville, Texas.

24. Menard County Historical Society (Ed.), *Menard County History: An Anthology*, (Anchor Publishing Company, San Angelo, Texas, 1982).

25. Mary Lewis Turner's oral presentation to the German-Texas Heritage Society at its September 5 -7, 1997 Convention at Kerrville, Texas.

26. Turner, M. L., *Julius Theodor Splittgerber*, Volume Two, pp. 212, 215.

27. Letter, Jacob Kuechler, Toyah, Pecos County (Texas) October 26, 1888, to Dearest Marie. Several individuals have transcribed the date of this letter as 1863. An example is in the book *Hermann Lungkwitz: Romantic Landscapist on the Texas Frontier* by James Patrick McGuire (University of Texas Press, Austin, Texas 1983), p. 20. This is incorrect. Toyal and Pecos County where not created until 1881 and 1883. Copy of letter provided by Anne Stewart.

17 – When Did the Insurgents Leave?

MYTH: The insurgents left the Hill Country for Mexico on August 1, 1862.[1]

FACT: The insurgents began gathering near Turtle Creek on July 31 and August 1, 1862. They departed from Turtle Creek on August 3, 1862, and arrived at the insurgent's base camp on the South Fork of the Guadalupe River the afternoon of August 3, 1862. They departed from their base camp on August 4, 1862.[2]

DISCUSSION: As shown above there were two locations where the insurgents gathered. One was on the South Fork of the Guadalupe River. This was a base camp for the individuals who had fled to the mountains earlier, mainly men from Gillespie County. The second was an assembly of men from Kerr and Kendall Counties.

In his book, Sansom says they left on August 1, 1862. However, this was the day Sansom, the two Degener brothers, and William Tellgmann left Comfort for Turtle Creek. [3] In a much later letter to James T. De Shields, dated August 14, 1907, Sansom stated it was August 3, when they departed from Turtle Creek. [4]

ENDNOTES - MYTH # 17

1. Sansom, *Battle of Nueces*, p. 4, Letter, Eduard Degener, August 1, 1862.

2. Sansom, *Battle of Nueces*, p. 4, Letter, John W. Sansom to James T. De Shields, August 14, 1907.

3. Sansom, *Battle of Nueces*, p. 4.

4. Letter, John W. Sansom to James T. De Shields, August 14, 1907.

5. Letter, Ernst Cramer to his parents, October 30, 1862.

18 – Betrayed By a Spy Among Them?

MYTH: Duff had a spy in the fleeing insurgent group who slowed down the movement and left signs for the pursuing Confederate force. [1]

FACT: The Confederates had no spy or informant in the fleeing group. The reason for traveling so slowly was that they had to stop at waterholes. To have travelled any further on each day would have resulted in a dry camp with no water for the men and horses. The group had a large number of horses and pack animals which experienced difficulties in the mountainous terrain.

DISCUSSION: From R. W. Williams' description of the route it can be determined where the insurgents stopped for the night. A careful study of the route taken reveals they were forced to stop at those sites where they made camp because to travel further would have resulted in a "dry camp" with no water for neither man nor animals. Even so, on August 8, they were forced to make a "dry camp" deep in the mountains. The next day they reached the West Prong of the Nueces River. Cramer explains, "Our horses being hungry and tired we decided to rest there for the day and the next morning to continue on our way to the Rio Grande." [2] Hoffmann says, "Then at the west fork of the Nueces, we halted at the waterhole for a day. Our horses being mostly unshod, had began to go lame on the flinty mountain trails." [3]

Several of the insurgents later wrote about the trip to the Nueces River. They all say they were travelling at a slow pace because they did not think the Confederates knew of their departure. John Sansom said, "Major Tegener while moving steadily on made no haste." Ernest Cramer said, "We had no suspicion of being betrayed or that we were being followed." Jacob Kuechler said, "We travelled very slow on account of the rough country one had to pass through, the heavy load our Pack animals had to carry." Helen Raley has August Hoffmann saying, "We were in no hurry. Sparing our horses and the few pack animals which carried our provisions, making camp in several places. We took more than a week for as much of the journey as might easily have been accomplished in three days." [4]

ENDNOTES - MYTH # 18

1. Smith, Thomas Tyree *Fort Inge: Sharps, Spurs, and Sabers on the Texas Frontier 1849-1869* (Eakins Press, Austin, Texas, 1993), p. 145.

2. Cramer's Letter.

3. Helen Raley's "Blackest Crime in History of Texas Warfare".

4. Sansom, *Battle of the Nueces*, p. 4; Cramer's Letter; Jacob Kuechler's Letter; Helen Raley, "Blackest Crime in History of Texas Warfare".

19 – Unarmed, Or Well-Armed?

MYTH The Unionists were poorly armed. One man had no weapon. [1]

FACT: The insurgents were all armed with muzzle loaders, shotguns, and pistols. [2]

DISCUSSION: The myth that the insurgents were poorly armed and one man had no weapons appears only in the Sansom's *Battle of the Nueces*. Another Nueces survivor, Henry Schwethelm, in an interview by Albert Schutze said, "At the appointed time 68 young men, the oldest not more than 35 years of age, fully equipped with rifles and six-shooters, the rifles mostly of German make (sic) mounted on good horses with pack animals, met at the designated place."

The Confederates captured 46 weapons. [3] An analysis of the weapons possessed by survivors shows that they had kept 46 weapons. This brings the total that the insurgents possessed to at least 92 weapons. This was a well-armed group.

ENDNOTES - MYTH # 19

1. Sansom, *Battle of the Nueces*, p. 10.

2. Albert Schutze, Was a Survivor of the Nueces Battle, p. 25.

3. McRae's Report, p. 615.

20 – Urged to Return, or Departed Voluntarily?

MYTH: On the afternoon of August 9, 1862, two strangers visited the insurgent's camp and encouraged them to return home as "things could be arranged, it would be all right to go back." Therefore about twenty-eight Unionist left camp in the afternoon or early evening of August 9, 1862.[1]

FACT: No one visited the insurgent's camp on August 9, 1862, or at any time after they left the base camp on the West Fork of the Guadalupe River.[2]

DISCUSSION: This is a perfect example of an interviewer misquoting the person being interviewed to make it seem as if they said something they did not say. Albert Schutze's interview with Henry Schwethelm says, "That night 28 of the German troop gave up the trip and returned to Fredericksburg and vicinity, by taking a new route from that by which the came, and managed to evade anyone in pursuit of them (some of those that left) were shot or hung (sic) later during the war."[3] A detailed study of this account shows Schutze was referring to those who "left early" on the night of August 9, 1862. That night of August 9-10 P.M., twenty-three insurgents left the camp after the opening shots were fired, at about three A.M. of August 10th, and before the Confederates launched a ground attack about six A.M. Several insurgents told about those leaving after three A.M. and before six A.M. Sansom says, "The camp was abandoned by able-bodied defenders."[4] Kuechler says, "After their end attempt to take our camp about half of our men could not stand the fight any longer and left."[5] Schwethelm says, "Some twenty-five men or so had left our camp before day (light) when the fight commenced."[6] Cramer says, "Men who had joined us at the

Guadeloupe (sic) deserted their posts one by one." [7] Siemering says, "During this time . . . fourteen Germans left the battle." [8]

Helen Raley's interview with August Hoffmann says, "At this point the survivor (Hoffmann) was questioned concerning some of the part who abandoned the enterprise, going back home. Capt. Henry Schwethelm who escaped after the battle, becoming an officer in the Federal army, stated before his death in 1924. *(An incorrect statement; Schwethelm was never a federal officer. After the war he was a Ranger captain, thus was called 'captain.')* Mr. Hoffmann agrees, thinking the number may have been higher, up to thirty-five. Soberly he picks up his narrative." [9] The Hoffmann interview addresses the statement that "two men had visited us in the camp and told them "things could be arranged – it would be all right to go back." This visit took place at either the Guadalupe River camp or the camp at Bushwhack Creek.

Schwethelm was clearly talking about those who "turned back the night of August 9-10." Hoffmann was clearly talking about those who "turned back at the Guadalupe base camp or at Bushwhack Creek." John Sansom wrote about those who turned back at the base camp. Sansom says, "Assembled there (Turtle Creek and West Prong Guadalupe River) and recognizing Major Tegener as their leader were about eighty men. In the afternoon of the following day, August 1st 1862, sixty-one of these, including myself and Major Tegener, set out for the Rio Grande." [10] If Sansom's numbers are correct – that eighty men were gathered and sixty-one left for the Rio Grande – then about nineteen insurgents turned back even before the group left the Hill Country.

ENDNOTES - MYTH # 20

1. Raley, "Blackest Crime in History of Texas Warfare".

2. Sansom, *Battle of the Nueces*, p. 4.

3. Albert Schutze, Was a Survivor of the Nueces Battle, p. 26.

4. Sansom, *Battle of the Nueces*, p. 11.

5. Jacob Kuechler's Letter.

6. Schwethelm's Letter of May 16, 1913.

7. Cramer's Letter of October 30, 1862.

8. Siemering, *Germans in Texas During the Civil War*, June 5, 1923.

9. Raley, "Blackest Crime in History of Texas Warfare".

10. Sansom, *Battle of the Nueces*, p. 4 and Weinheimer, Ophelia Nielsen, *The Early History of Gillespie County, Texas*, (Thesis, Southwest Texas State Teachers College, August 1952), p. 61.

21 – A Bloody and Unprovoked Massacre?

MYTH: Lieutenant Colin D. McRae, who was the commander of the pursuit force, ambushed and "rode down upon the (insurgent) camp while the Germans lay sleeping. He surrounded it and opened fire indiscriminately." The Confederates then "rode into their (the insurgent) midst, firing point blank and slashing with cavalry sabers." "The result was a massacre. Nineteen Germans were killed by gunfire, and six more trampled to death by McRae's cavalry. Nine surrendered. McRae ordered them shot and they were executed on the spot." [1]

FACT: McRae made a dismounted attack on the camp. The type of military maneuver he used is termed a double envelopment. [2]

DISCUSSION: Every element of this particular myth is incorrect. The Confederate force did not ride "down upon the camp while the Germans lay sleeping. McRae did surround the camp. Nor did he open fire indiscriminately. The Confederates did not ride in the midst of the insurgents, firing point-blank and slashing with sabers. The Confederates made a dismounted attack about six o'clock in the morning of August 10th. The result was not a massacre. The insurgents put up a stubborn defense, forcing the Confederates to retreat three times. It is correct that nineteen insurgents were killed at or near the battle site, but no defender was trampled to death by McRae's cavalry, and no insurgent surrendered; those wounded who had volunteered to cover the withdrawal of their comrades fought until overwhelmed and overrun. Several of those wounded were captured alive, but they were not executed on the spot. The fighting was over at about nine in the morning. The wounded insurgents were "treated the best they could have," since the Confederates had no surgeon. It is a fact the wounded insurgents were executed later in the day, about four in the afternoon.

Several descendants of the insurgents claim the Confederate ambushed the insurgents on the Nueces River. [3] Webster defines an ambushed as "a trap in which concealed persons lie in wait (emphasis added) to attack by surprise." [4] Webster defines envelopment as "the circumstances, objects, or conditions by which one is surrounded." [5] Thus a double envelopment is directed against or around both the enemy's flanks. [5]

This is a perfect example of individuals without military training using a word that is incorrect. The word 'ambushed' described a more violent military maneuver and thus is designed to distill more emotional feeling. The definition of ambush is to lie in wait. This is not what took place on the Nueces River. The insurgents were already deployed when the Confederates arrived. The insurgents consisted of sixty-nine men while the Confederates had a force of only ninety-two or ninety-three men. The normal ratio needed to attack a defensive position is 3:1 and even then some type of surprise or other faction is required. The Confederates did not have sufficient force to attack straight on. They needed a well-coordinated surprise attack. This was what McRae's plan was, but when Leopold Bauer discovered the Confederate force this plan was compromised. The Confederates attacked three times but were driven back each time. It was only when those twenty-eight insurgents "left early" and others had run out of ammunition and left, did the Confederate force had sufficient strength to overrun the camp.

Nineteen insurgents were killed at or near the battles site and this include about seven who were wounded seriously. It is a fact that those seven wounded were very badly injured, considered in combination with the fact that the Confederates did not have the manpower to carry them out with their own severely wounded. At about four P.M. August 10, 1862, they were executed.

ENDNOTES - MYTH #21

1. Fehrenbach, T. R., *Lone Star: A History of Texas and the Texans*, (Wings Books, New York, New York, 1968), p. 364.

2. Williams, *With the Borden* Ruffians, pp. 245-246

3. Several discussions with Gregory Krauter of Comfort, Texas, in 1994-1996

4. *Webster's New Collegiate Dictionary*, G & C. Merriam Company, Springfield, Massachusetts 1979, p. 36.

5. Dupuy R. Ernest, and Dupuy, Trevor N., *Military Heritage of America*, (McGraw-Hill, New York, New York, 1956), p. 20.

22 – Battle Casualties Or Cold-Blooded Executions?

MYTH: The insurgent casualties were nineteen killed, six trampled to death, and nine surrendered and afterwrwards executed. [1]

FACT: Nineteen insurgents were killed by gunfire or executed during or after the battle. [2]

DISCUSSION: The only source of a correct figure for casualties among the Nueces insurgents is the Treue der Union monument in Comfort, Texas, and even this is confusing. All other sources are greatly exaggerated. For example Williams says, "Sixty were killed by gunfire and twenty wounded, that were executed. [3] August Siemering says fifteen insurgents were left on the battlefield, "some were badly wounded. [4] Biggers says, "Nineteen of the refugees were killed and nine were wounded." [5] Schwethelm is quoted as saying, "Twenty-three had been killed or wounded." [6] Gilbert Benjamin in his study, *The Germans In Texas* claims "thirty-two were killed" and some were taken prisoner and later shot. [7] Thomas Smith in his book *Fort Inge* says the insurgents had "nineteen killed, nine wounded, and six run down by cavalry as they attempted to flee on foot." [8] Pirtle and Cusack in their book *Fort Clark* repeat the most common belief when they say, "Nineteen Germans lay dead … and nine wounded." [9] William Banta and J. W. Caldwell in their book *Twenty-seven Years on the Teas Frontier* claim that "fifty-two Germans were killed." [10]

Texas academics have even gotten into the casualty debate and none of them have the correct numbers. Some examples are: Robert Patttison Felgar's dissertation *Texas in the War for Southern Independence 1861-1865*, which claims "Nineteen killed and nine wounded . . . The wounded prisoners were murdered." [11] Ada Maria Hall's thesis *The Texas Germans in State and National Politics, 1850-1865*, reports, "Nineteen of the fugitives had been killed, and nine others were wounded." [12] Sara Kay Curtis' thesis *A History of Gillespie County, Texas, 1846-1900* says, "Nineteen of the Germans were killed and nine wounded," and, "The wounded were shot." [13] Frank W. Heintzen's thesis *Fredericksburg, Texas During the Civil War and Reconstruction* does not give the number of insurgents killed but it does state, "About 20 wounded (insurgents) were captured after the fight." [14] Ophelia Nielsen Weinheimer's thesis *The Early History of Gillespie County, Texas,* states, "Nineteen were killed in battle," and, "Nine were killed by Duff at the conclusion of the fight." [15] Melvin C. Johnson's thesis *A New Perspective for the Antebellum and Civil War Texas Germany Community* also does not specify the number of insurgents killed, but like Weinheimer, says, "The nine wounded Germans were murdered later in the day at the direction of a Lieutenant Luck (Lilly)"[16] and while Shawn Henderson' thesis *A Descriptive Case Study of the Encounter at the Nueces River on August 10, 1862* does not give the number of insurgents killed, but does report, "Eleven wounded German prisoners were being cared for." [17]

The basic reason for such inaccuracies and disparities, aside from poor research, is because of the way the Treue der Union monument at Comfort, Texas, records those casualties. It lists nineteen names as *'Gefallen am 10 August 1862 am Nueces'* (Killed at the Nueces August 10, 1862). Elsewhere on the monument nine names are listed as *'Gefangen genommend und emordet'* (Captured and murdered). It has been a common misconception since the monument was built; an assumption that those nine captured and murdered were the same ones captured and executed at the Nueces River. This is not correct! Those nine were 'later' captured and murdered were part of the twenty-eight who "left early" and were captured and executed in other parts of Texas. August Siemering, an original member of the Union Loyal League, tells how some of these nine were killed. He explains how the Confederates were, "successful in taking several stragglers who were on their way home. They were dealt with in a barbaric manner. Two of them (Stieler and Bruckisch), who were near their homes, weak and exhausted, were cruelly shot. Two brothers (Louis and Adolph Ruebsamen), who were wounded met the same fate. Two Fredericksburg men (Conrad Bock and Fritz Taps) were hanged in Boerne.

The correct number of insurgent casualties is nineteen, at or near the battle site – including those wounded executed by or at the orders of Lieutenant Lilly.

Wm. Paul Burrier

ENDNOTES - MYTH # 22

1. Fehrenbach, *Lone Star*, p. 364

2. Treue Der Union Monument, Comfort, Texas.

3. Williams, *With the Border Ruffians*, p. 248.

4. Siemering, *Germans in Texas During the Civil War*, June 5, 1923.

5. Biggers, *German Pioneers in Texas*, p. 58.

6. Schutze, Was a Survivor of the Nueces Battle.

7. Benjamin, Gilbert Giddings, *The Germans In Texas: A study in Immigration*, (Jenkins Publishing Company, Austin, Texas, 1974), p. 110.

8. Smith, *Fort Inge*, p. 144.

9. Pirtle III, Caleb and Cusack, Michael F., *Fort Clark: The Lonely Sentinel On Texas's Western Frontier*, (Eakin Press, Austin, Texas, 1985), p. 50.

10. Banta, Captain William and Caldwell, Jr., J. W., *Twenty-Seven Years on the Texas Frontier* (L. G. Park, Council Hill, Oklahoma, 1933), p. 186.

11. Felger, Robert Pattison, *Texas In The War For Southern Independence 1861-1965*, (Dissertation, University of Texas, Austin, Texas, 1947), p. 345.

12. Hall, Ada Maria, *The Texas Germans in State and National Politics, 1850-1865*, (Thesis, University of Texas, Austin, Texas, 1938), p. 92.

13. Curtis, Sara Kay, *A History of Gillespie County, Texas, 1846-1900*, (Thesis, University of Texas, Austin, Texas, 1943), p. 59.

14. Heintzen, Frank W., *Fredericksburg, Texas Suring the Civil War and Reconstruction*, (Thesis, St. Mary's University, San Antonio, Texas, 1944), p. 53.

15. Weinheimer, Ophelia Nielsen, *The Early History of Gillespie County, Texas*, (Thesis, Southwest Texas State Teachers College, San Marcos, Texas, 1952), p. 63.

16. Johnson, Melvin C., *A New Perspective for the Antebellum and Civil War Texas German Community*, (Thesis, Stephen F. Austin State University, Nacogdoches, Texas, 1993), p. 202.

17. Henderson, Shawn, *A Descriptive Case Study of the Encounter at the Nueces River on August 10, 1862*, (Thesis, Urbana University, Urbana, Ohio, 2007), p. 19.

23 – Sworn to Take No Prisoners?

MYTH: The reason that the Germans were executed was because Duff either ordered or stated he wanted no prisoners. It was Duff's men under the command of one of his officers, a Lieutenant Luck (Lilly) who took command after McRae was wounded, and who murdered those who had surrendered. [1]

FACT: Neither Captain Duff nor any other Confederate command authority ordered the wounded insurgents executed.

DISCUSSION: Despite numerous books and articles which claim Duff was in command of the Confederate forces at Fredericksburg, he was not. Confederate Brigadier General H. P. Bee ordered two battalion task forces to the Texas Hill Country in late July and early August of 1862. One was commanded by Lieutenant Colonel Nathaniel Benton and consisted of four companies of the 32nd (aka 36th) Regiment Texas Cavalry. [2] It was sent to the northern counties of Llano, Burnet, and Mason. The second task force was under the command of Captain John Donelson and consisted of Company K, 2nd Regiment Texas Mounted Rifles, James Duff's Company of Partisan Rangers, and three companies or detachments from Joseph Taylor's 8th Battalion Texas Cavalry. It was sent to Gillespie, Kerr, and Kendall Counties. [3] General Bee's instructions were, "To issue a proclamation declaring martial law, and requiring all good and loyal citizens to return quietly to their homes, and take the oath of allegiance to the Confederate and State governments, or be treated summarily as traitors in arms." Bee further ordered the two task forces to "send out scouting parties into the mountain districts with orders to find and break up any encampments and depots as had been report to exist there, and to send the families and provisions back to the settlements." [4]

Colonel Benton was appointed both provost marshal and troop commander of the northern task force. In the south task force this duty was split. Donelson was appointed the troop commander while James Duff was appointed the provost marshal. Duff did not have command of any of the troops sent to Fredericksburg, Kendall or Kerr Counties. [5]

Donelson established his camp on the Pedernales River west of Fredericksburg. James Duff established the provost marshal office at Fort Martin Scott, just east of Fredericksburg. Duff was never stationed at Camp Pedernales. It was Donelson, not Duff who ordered the pursuit force to overtake the insurgents. [6]

The Confederates overran the insurgent's camp by nine A.M. They treated their own wounded and those of the insurgents as best as they could as they had no doctor. [7] If there had been orders to "take no prisoners" as many claim, the prisoners would have been quickly executed. By waiting some seven hours after overrunning the insurgent camp argues that the Confederates were not operating under a 'no prisoner' order. As an infantry officer with two intensive combat tours in Vietnam, one in the Dominican Republic, and one in Granada, I can state unequivocally that if the prisoners were to be executed they would have been so immediately – before any relationship developed between the two groups, captives and captors. The allegation that Lieutenant Lilly took command after Lieutenant McRae was wounded is ludicrous; another example of someone without military training or study of military science writing about a purely military incident. Of five Confederate officers present, Lilly was the junior in rank. There were three other first lieutenants and another second lieutenant senior to Lilly. No junior officer would have taken it upon himself to commit such a dastardly deed. It is, however, a military tradition that the junior officer is given the most unpleasant task. What this suggests in this circumstance is that the decision to execute the prisoners was a deliberate one. The most logical reason for such a horrific decision was that McRae's command was overtaxed with caring for their own wounded, captured equipment and animals, and did not have the necessary means to carry out the wounded insurgents. An examination of the numbers clarifies McRae's dilemma. McRae's force numbered ninety-six men. He had two killed and nineteen wounded of his force, leaving him with seventy-three effective men. Using McRae's list of causalities and cross checking it with the service records of each Confederate wounded, it can be determined that at least eight Confederates had to be carried out. The decision to use hand litters with four men per litter would have meant that as least thirty-two men would be required to move the Confederate wounded any distance. That reduced the effective force down to forty-one. If he transported another seven wounded (the likely number of insurgent wounded) his force would have been reduced by another twenty-eight litter-bearers, leaving only fourteen. How were fourteen men going to lead one hundred seventy-eight animals, many loaded down with captured equipment?

Most likely McRae had a council of his officers to decide what to do. The decision was made to execute the wounded insurgents. The junior officer, Lilly, was given the task. A member of the insurgent force, writing years after the war, said that Lilly, "called for volunteers to kill them and he got plenty of them." [8] One final point of this discussion would be that both most writers and historians still claim it was Duff's company that killed the wounded. Again, I would reiterate that this is not correct. R. H. Williams, a member of Duff's company, "denounced the bloody deed in as strong language as I could use, telling the perpetrator, (Lieutenant Lilly) to his face, what he was, and what every decent, honourable (sic) man would think of him as long as they lived. He handled his six-shooter, and looked as though he would like to use it on me; but the coward was afraid to shoot at a live man, as I told him. Fortunately some of my own comrades backed me up, or I have no doubt it would have gone hard with me." [9] Williams' statement makes it clear it was not Duff's men who executed the wounded insurgents. Instead they were vehemently opposed to the executions.

Two names of some of those who executed the insurgents are known. Both were members of Davis' company.

ENDNOTES - MYTH #23

1. Williams, *With the Border Ruffians*, pp. 236, 250 & 259; Biggers, *German Pioneers*, p. 59; Smith, *Fort Inge*, 1993, p. 145; McGowan, Stanley S., *Horse Sweat and Powder Smoke: The First Texas Cavalry in the Civil War*, Texas A & M Press, College Station, Texas 1999, p. 70; Edwards, *The Story of Fredericksburg* ,pp. 35-36; Pirtle III, Caleb, Cusack, Michael F.. *Fort Clark: The Lonely Sentinel On Texas's Western Frontier*, pp. 50-51; and Simpson, Colonel Harold B., & Wright, Brigadier General Marcus J., *Texas In The War* 1861-1865 (Hill Junior College Press, 1965), p. 145.

2. Report, Lieutenant Colonel Nat Benton's, *San Antonio Herald*, August 30, 1862.

3. Report, Brigadier General H. P. Bee, Headquarters Sub-Military District of the Rio Grande, San Antonio, Texas , October 21, 1862, to Headquarters First District of Texas, San Antonio, Texas Official Records, Series I, Volume LIII, pp. 454-456.

4. Ibid.

5. Ibid.

6. Camp Pedernales Post Returns for the month of August 1862, Document # 292, Record Group, 109, National Archives, Washington, D. C.

7. Williams' *With the Border Ruffians*, p. 246.

8. Schwethelm's Letter.

9. Williams, *With The Border Ruffians*, p. 250.

24 – Outnumbered and Ready to Surrender?

MYTH: The Confederate force greatly outnumbered the Unionists. Therefore there was no need to open fire on the Unionist. They would have surrendered if asked. [1]

FACT: The Confederates did not have sufficient force to take the camp. Therefore, they needed a sound and well-executed tactical maneuver. Surprise is one of the principles of war. [2] As George S. Patton said, "The goal is not to give your life for your country; it is to make the other SOB give his life for his country." [3]

DISCUSSION: Eduard Degener, the leader of the Union Loyal League when talking about asking the insurgents to surrender said, "He knew the men well, that they were the best shots in the country, and that they would never surrendered." [4] Degener asked W. J. Edwards, a private in Duff's Company, "If we asked them to surrender? I told him we did not, telling him we supposed they would not (have) surrender(ed), from what we heard." [5]

No military commander who is about to attack, stops and tells the enemy they are there and asks them to surrender. It would be foolish to alert the enemy of an impending attack. The enemy would be fully alert and have the best defensive positions as possible.

ENDNOTES - MYTH #24

1. Sansom, *Battle of Nueces*, pp. 10-11; Biggers, *German Pioneers*, p. 59; Raley, Blackest Crime in Texas Warfare; Handout, Comfort Foundation; and Ransleben, Guido E. , *A Hundred Years of Comfort*, p. 91.

2. Dupuy, *Military Heritage of America*, p. 8.

3. Ruggero, Ed, *Combat Jump: The Young Men Who Led the Assault into Fortress Europe*, Harper Collins, New York, New York, 2003, p. 120.

4. Degener's Letter.

5. Records of the Confederate Military Commission in San Antonio July 2-October 10, 1862." Edited by Alwyn Barr, *Southwest Historical Quarterly*, Volume LXXIII (October 1969), p. 2.

25 – Casualties Among the Confederates?

MYTH: The Confederate casualties numbered ten to twelve and with eighteen to forty-five wounded. About half of the wounded later died at Fort Clark. [1]

FACTS: McRae's forces suffered two killed and nineteen wounded, of whom four later died. [2]

DISCUSSION: The numbers of Confederates killed and wounded have been greatly exaggerated by many writers and historians. One of the major themes is the insurgents were poorly armed, surprised and massacred while at the same time inflicting many Confederate casualties. An example of this theme is Sara Kay Curtis in her thesis, which says, "Duff's men attacked this almost unarmed, defenseless band of Germans." While at same time, "A total of forty-seven Confederates were killed." [3]

Don Biggers in his book *German Pioneers* says, "During the battle following Duff's attack, 12 confederates were killed and 18 were wounded." [4] Albert Schutze in his article, "Was a Survivor of the Nueces Battle," claims Schwethelm "learned from Doctor Downs that Duff's company had lost 10 men killed, (with) 46 wounded and afterwards learned from the same source that half of the wounded died after they reached Fort Clark." [5]

Even some Confederate accounts exaggerate their causalities. For example, R. H. Williams, a private in Duff's Company, says, "a loss on our side of twelve killed and eight wounded." [6] Of the eight wounded Williams claims five Confederates died at Fort Clark. [7]

ENDNOTES - MYTH #25

1. Williams, *With the Border Ruffians*, p. 248; Biggers, *German Pioneers*, pp. 59-60; and Schutze, "Was a Survivor of the Nueces Battle," p. 27.

2. McRae's Report, p. 615, McRae's Casualty List and Compiled Service Records of the individuals wounded.

3. Curtis, Sara Kay, *History of Gillespie County, Texas, 1846-1900*, University of Texas, Austin, Texas, 1943.

4. Biggers' *German Pioneers*, pp. 59-60.

5. Schutze, "Was a Survivor of the Nueces Battle."

6. Williams, *With The Border Ruffians*, p. 248.

7. Ibid., p. 254.

26 – Executed Out Of Hand?

MYTH: Any member of the Unionist group captured later was executed immediately. [1]

FACT: At least three survivors of the Nueces battle were later captured but not executed.

DISCUSSION: These three men included Ferdinand Simon, William F. Klier, and August Duecker. There is a fourth man, William Vater, who may have also been captured but not executed.

Ferdinand Simon was seriously wounded at the Nueces. His comrades carried him safety away from the battle site, but later he became separated from the survivor's group and a Confederate scout captured him about four days after the battle and took him to Fort Clerk. [2] Simon was tried by the Confederate Military Commission and sentenced to death. [3] Before the execution could be carried out, martial law was annulled and his life spared. [4] He was transported to Austin for trial. It appears he spent the rest of the war in jail. Ferdinand Simon died in July, 1878, at his home near Boerne. [5]

William F. Klier survived the Nueces battle and returned to Gillespie County. [6] The Confederates arrested him. His brother-in-law, Julius Ransleben, convinced the Confederates not to hang Klier. He was transported to Boerne where he was forced to watch the hanging of Conrad Bock and Fritz Tays. [7] Klier was conscripted into Company B, 3rd Regiment Texas Infantry and served until the end of the war. [8] William F. Klier died on June 2, 1907 in Gillespie County. [9]

August Duecker survived the Nueces battle and returned to his home in Gillespie County where he hid in his attic. [10] His wife died January 1, 1863, while he was in hiding. [11] When he came out of hiding he agreed to serve as a Confederate teamster. Captain Van der Stucken arrested him and he was held in the Austin jail. Oral accounts claim that he escaped and hid for the remainder of the war. [12] August Duecker died on April 19, 1894, in Gillespie County. [13]

The fourth man, William Vater was seriously wounded at the Nueces. He hid in the river. After the battle the insurgents treated him as best they could, but were forced to leave him at the site. He was rescued by Callimense Beckett from Uvalde, Texas, who took him to Brackettville for medical treatment. He was later taken to San Antonio. [14] No further reference is made of William Vater.

ENDNOTES - MYTH # 26

1. Sansom, *Battle of Nueces*, p. 11.

2. McRae's Report, p. 615

3. Records of the Confederate Military Commission in San Antonio, July 2-October 1862, Edited by Alwyn Bar in Southwester Historical Quarterly , Volume LXXIII, No. 2, October 1869, pp. 270-272.

4. Letter and General Order Number 66, Headquarters Confederate Army Richmond, Virginia, September 12, 1862, contained in OR, Series I, Volume IX, pp. 735-736.

5. Genealogical Abstractions From Kendall County, Texas Probate Records 28 April 1862-10 December 1900, Published by Genealogy Society of Kendall County, Texas, 1886, pp. 195-197.

6. Striegler, Selm, *The Striegler Family History*, n. d, pp. 77-78.

7. Schlickum's Letter.

8. William Kleir's Confederate Records.

9. Records of Der Friedhof Cemetery, Fredericksburg, Texas, p. 44.

10. Hoffmann's Letter.

11. Records of Der Friedhof Cemetery, Fredericksburg, Texas, p. 151.

12. Oral interview with Amelia "Mollie" Eckhardt Dennis, a descendent of August Duecker on September 5, 1997.

13. Records of Der Friedhof Cemetery, Fredericksburg, Texas, p. 16.

14. Hoffmann's Letter, Kuechler's Letter and Anthon, Florence, (complier) *Early History of Uvalde and Surrounding Territory*, (West Main Press, Uvalde, Texas 2006), p. 103-105.

27 – The Journey of the Stieler Women

MYTH: The mother and sister of sixteen-year-old Heinrich Stieler travelled to the Nueces River and recovered his body and that of his friend Theodor Bruckisch. [1]

FACT: Heinrich Stieler and Theodor Bruckisch were not killed on the Nueces River. Stieler was captured and murdered at Goat Creek in Kerr County. [2] It is very likely Bruckisch was also killed at Goat Creek.

DISCUSSION: According to the myth, upon hearing of the death of Heinrich Stieler his mother, forty-year-old Wilhelmina Urban Stieler, and his sister, seventeen-year-old Wilhelmina Stieler, fearful that his body would be prey to the buzzards, decided to ride horses to "the scene of slaughter" and recover the body. [3] The myth claims that Heinrich Stieler and his friend eighteen-year-old Theodor Bruckisch were killed at the same time. [4]

From the time this compiler began researching the events of the Nueces River Battle and Massacre, this story just did not ring true. The myth claims that the two insurgents "had escaped from the Nueces battleground" [5] and "afterward fell into the hands of the Confederates." [6] If they had "escaped" from the battle site and "afterward" fell into the hands of Confederates, then they could not have been killed on or near the Nueces River. The myth goes on to claim that when captured, Stieler had just about convinced the Confederates he was on his way to join the Confederate Army when Bruckisch was captured. [7] If Stieler was killed at or near the Nueces River – far from the settlements – he could not have convincingly claimed that, "No he was on his way to join the Confederates." The only individuals living in the area were a few settlers and Indians, none of whom were recruiting for the Confederates. The myth claims that when Bruckisch was captured he "betrayed himself and Stieler." [8] The two were shot immediately.

One of the insurgent's descendants, in reference to Heinrich Stieler's mother and sister, described the two ladies as having, "Packed some lunch and left early in the morning." [9] If the two ladies went to the battle site to recover the bodies, they would have had to pack more than just a lunch. Remember; it took both the fleeing insurgents and Confederates almost a week to reach the Nueces River.

Another part of the myth says the women, "Passed through the ranks of the soldiers, who, learning of (their) mission, were entirely chivalrous and did not molest (them) or prevent (them) from carrying out their intention." [10] Again, this could not have been the case; there were no Confederate troops at or near the battle site who could have molested them.

The myth goes on to claim that after travelling over a hundred miles from their homes to the battle site, the two women discovered that burying the two bodies would overtax their strength. So they gathered brush with which they covered the corpses, weighting them down with rocks so the wild beasts could not tear away the cover. The two women then returned to their home. In the course of time, they returned again to the spot and brought their remains of young Stieler and Bruckisch to their final resting place on their farm. A journey of that magnitude – twice – does not meet the common sense and logic test. It took the fleeing group over a week to travel from Comfort to the Nueces River. To believe this part of the myth one would have to believe the two ladies travelled alone, over a hundred miles in Indian country – on two occasions. This might make a wonderful legend but it is too far-fetched as to be possible. This is not intended to take anything away from the bravery of the two women. There is no doubt that they did go to the "the scene of slaughter." It is just the scene was not at or near the Nueces River.

The correct facts can be established by a little detailed research into the event. August Siemering, a major member of the Organization provides some information on Stieler and Bruckisch. He says, "(They) had come within reach of their homes and were found thirsty and tired and were shot without pity." [11]

Fritz Schellhase, a member of the fleeing insurgent group recalls a story he heard. They fell out with each other. One said, "We have to go right." The other said, "We have to go left." But Stieler knew he was right. He went down the river until he got to the Kerrville road and reached the home of Sidney Rees. He knew he needed a travel permit and asked Rees, who was the Kerr County clerk, to help him get one. Rees told him since he was not yet 18 years old he did not need one. Stieler next stopped at Mrs. Fritz Tegener house. As he came into her house, Mrs. Tegener said, "That's my Fritz's gun, how did you get it?" Stieler replied, "I traded with your Fritz." She asked, "Where is Fritz, did he come through?" Stieler replied, "I don't know, me and Fritz were separated and were not together for some time." At that time Daniel Boone Lowrance arrived with Bruckisch. At the same time, a Confederate scout arrived. Lowrance turned Bruckisch over to him. The soldiers asked Bruckisch if Stieler was in the battle too. He answered, 'yes.' [12]

Henry J. Schwethelm, another member of the August insurgent group, tells of Stieler's and Bruckisch's death in a 1913 letter. He says, "Stieler and Bruckisch were taken prisoners near Kerrville by James Starkey, the Kerr County Provost Marshal, and turned them over to some of Captain Davis company about 3 miles above Kerrville on Goat Creek. They were used as shooting targets." [13]

One of the men who shot Stieler or Bruckisch was J. M. Seal of Davis' company. He related in a September, 1862, letter, "I had the pleasure of shooting one of the poor devils a few days ago myself." [14]

What can be determined is the data contained in the myth is somewhat correct. It is because the individuals who later wrote about the event did not relate this portion of the story in detail. Wilhelmina Urban Stieler and Wilhelmina Stieler did not go all the way to the Nueces River to retrieve the bodies. But they did go to the "scene of slaughter" – at Goat Creek, rather than the Nueces River.

ENDNOTES - MYTH # 27

1. Ransleben, Guido E., *A Hundred Years of Comfort in Texas*. (The Naylor Company, San Antonio, Texas 1954), p. 95.

2. Stewart, Anne and Mike, *Comfort Women in Comfort History*, (Privately published by Anne and Mike Stewart, Comfort, Texas. 1993), pp. 128 – 130.

3. Ransleben, *Hundred Years*, p. 95.

4. Ibid.

5. Dykes-Hoffman, Judith, *Treue Der Union: German Texan Women On The Civil War Homefront*, (M.A. thesis, Southwest Texas State University, San Marcos, Texas, 1966), pp. 81-82.

6. Ransleben, *Hundred Years*, p. 95.

7. Ibid.

8. Ibid.

9. This descendent is Tillie Reeh Heinen Lott. Her account is contained in Anne and Mike Steward's *Death on the Nueces*, pp. 36 – 37.

10. *Death on the Nueces: Mina Stieler, Stories by Anne and Mike Steward,* (Privately published by Anne and Mike Seward, Radium Springs, New Mexico, 1997), pp. 24-41.

11. Siemering, August, *Germans in Texas During The Civil War*, San Antonio Freir Presse fuer Texas, June 8, 1923.

12. Schellhase Warren W., "Edited Journal of Fritz Schellhase", n. d., Copy in possession of author.

13. Letter, Henry Schwethelm To His Grandson, Otto, Kerrville, Kerr County, May 16, 1913. Copy in possession of author.

14. Letter, J. M. Seal, Camp Davis to Dear Friends, September 8, 1862. Copy in possession of author.

BIBLIOGRAPHY

GOVERNMENT DOCUMENTS

An Act to Alter and Amend an Act Entitled "An Act For The Sequestration of the Estates, Property, and Effects of Alien Enemies and for Indemnity of Citizens of the Confederate States, and Persons Aiding the Same in the War With The United States', August 13, 1861, OR, Series II, Volume II.

An Act to Provide for the Protection of the Frontier of the State of Texas, passed by the Texas Legislature on, February 7, 1862, Laws of Texas, 1822—1897, Compiled by H. P. N Gammel, Gammel Book Company, Austin, Texas, 1898.

An Act to Provide For the Protection of the Frontier of the State of Texas, passed by the Texas Legislature on December 21, 1861, Laws of Texas, 1822—1897, Compiled by H. P. N. Gammel, Gammel Book Company, Austin, Texas, 1898.

An Act Respecting Alien Enemies', approved August 8, 1861, Official Records (OR), Series II, Volume II.

Camp Pedernales Post Return for the Month of August, 1862, National Archives, Washington, D.C.

Camp Pedernales Post Return for the Month of September, 1862, National Archives, Washington, D.C.

Case 101, Gillespie County, Texas District Court Records.

Case Number 5, Kendall County, Texas District Court Records.

Charles Burgmann's Confederate Service Records, Record Group 109, National Archives, Washington, D.C.

Deed Records, Gillespie County, Texas.

Deed Records, Kendall County, Texas.

General Order Number 1, Headquarters Frontier Regiment, Texas Rangers, Austin, February 1, 1862, Adjutant Generals Correspondent, Texas State Archives, Austin, Texas.

General Order No. 8, Headquarters, Troops in Texas, San Antonio, May 24, 1861. *Austin State Gazette*, June 8, 1861.

General Order Number 9, Headquarters 7th Military Department, Fort Smith, October 23, 1849, Records of the Adjutant General's Office (Record Group 94), National Archives, Washington, D.C.

General Order Number 45, Headquarters, Department of Texas, Houston, Texas, May 30, 1862, *War of the Rebellion: A Compilation of the Official Records of the Union and Confederate Armies*, (OR), Series I, Volume 9.

General Order Number 66, Headquarters Confederate Army, Richmond, Virginia, September 12, 1862, OR, Series I, Vol. IX.

Eighth U. S. Census, 1860 Bexar Census, p. 3b.

Kerr County, Texas Probate Book A.

Kerr County, Texas Naturalizations Records.

Letter. Michael E. Pilgrim, Textual Reference Division, National Archives, Washington, D.C., January 31, 1995, containing a copy of Lieutenant C. D. McRae's Casualty List, Records Group 109, E22.

"Minutes of Meeting, Gillespie Rifles, February 23, 1862, and March 29, 1862, with copy of the Gillespie County Rifles Resolution, District Clerk's Office, Fredericksburg, Texas.

Muster Roll, Captain William A. Blackwell's Company, May 4, 1861, Texas State Archives, Austin, Texas.

Muster Rolls, Captain Philip Braubach's Company, February 7, 1861, May 25, 1861, August 25, 1861, November 25, 1861, and February 25, 1862, Texas State Archives, Austin, Texas.

Muster Roll, Captain George Freeman's Company, March 1, 1861, & November 18, 1861, Texas State Archives, Austin, Texas.

Muster Roll, Captain William T. Harbour's Company March 5, 1861, Texas State Archives, Austin, Texas.

Muster Rolls, Company K, 5th U. S. Infantry, December 31, 1849— June 13, 1851, Records of the Adjutant General's Office (Record Group 94), Washington D.C.

Muster Rolls, Company K, 5th U. S. Infantry, August 31, 1853— February 28, 1854.

Muster Roll, Captain Charles H. Nimitz's Company, July 31, 1861, Texas State Archives, Austin, Texas.

Orders Number 10, Headquarters, 7th Military Depart, Fort Smith, June 9, 1849, Records of the Adjutant General's Office (Record Group 94 National Archives, Washington, D.C.

Proclamation by Jefferson Davis, President Confederate States of America, August 14, 1861, OR Series II, Volume II.

Records, 31st Brigade District, Texas State Troops, A.G.C., TSA, Austin, Texas and Quarterly Returns, 31st Brigade, July 1, 1862, & October 1, 1862, A.G.C., TSA, Austin, Texas.

Report, Brigadier General H. P. Bee, Headquarters Sub-Military District of the Rio Grande, San Antonio, Texas October 21, 1862, to Headquarters First District of Texas, San Antonio, Texas OR, Series I, Volume LIII.

Report, Brigadier General P.O. Hebert, Headquarters First District of Texas, San Antonio, Texas, October 11, 1862, to General Cooper, Adjutant-General, Richmond, Virginia, OR, Series I, Volume LII.

Report, Captain H. T. Davis, to Headquarters Texas Frontier Regiment, July 25, 1862, A.G.C., TSA.

Report, Headquarters 31st Brigade, Texas State Troops, New Braunfels, Comal County, March 8, 1862, A.G.C., TSA, Austin, Texas.

Report, James Duff, Captain, Commanding Company of Partisan Ranger, Headquarters Camp Bee, San Antonio, Texas, June 23, 1862, OR Series I, Volume 9.

Texas Governor's Proclamation Book, TSA, Austin, Texas.

William Kleir's Confederate Service Records, Record Group 109, National Archives, Washington. D.C.

BOOKS

Anthon, Florence, Compiler, *Early History of Uvalde and Surrounding Territory*, West Main Press, Uvalde, Texas, 2006.

Banta, Captain William and J. W. Caldwell, Jr., *Twenty-Seven Years on the Texas Frontier*, Published by L. G. Park, Council Hill, Oklahoma, 1933.

Baum, Dale, *The Shattering of Texas Unionism: Politics in the Lone Star State During the Civil War Era*, Louisiana State University Press, Baton Rouge, Louisiana, 1998.

Benjamin, Gilbert Giddings, *The Germans In Texas: A Study in Immigration*, Jenkins Publishing Company, Austin, Texas 1974.

Bennett, Bob, *Kerr County Texas 1856 – 1956*, The Naylor Company, San Antonio, Texas, 1956.

Biesele, Rudolph Leopold, *The History of the German Settlements in Texas 1831-1861*, Eakin Press, Austin, Texas, 1986.

Biggers, Don H., *Germans Pioneers In Texas*, Fredericksburg Standard, Fredericksburg, Texas, 1925.

Bitton, Davis, Editor, *Reminiscences and Civil War Letters of Levi Wight*, University of Utah Press, Salt Lake City, Utah, 1970.

Buenger, Walter L., *Secession and the Union in Texas*. The University of Texas Press, Austin, Texas, 1984.

Dupuy, R. Ernest and Dupuy, Trevor N., *Military Heritage of America*, McGraw-Hill, New York, New York 1956.

Edwards, Walter F., *The Story of Fredericksburg, Its past, present, points of interest and annual events*, Fredericksburg Chambers of Commerce, Fredericksburg, Texas, 1969.

Edwards, Walter F., *Tales of Old Fredericksburg*, Published by Walter Edwards, Fredericksburg, Texas 1975

Fehrenbach, T. R., Lone Star: A History of Texas, American Legacy Press, New York, New York, 1983.

Francis, Mary E., *The Hermit of the Cavern*, Naylor Printing Company, San Antonio, Texas, 1932.

Genealogy Society of Kendall County, Texas, Genealogical Abstraction From Kendall County, Texas Probate Records 28 April—10 December 1900, Published by Genealogy Society of Kendall County, Texas 1986.

Glenn, Frankie Davis, *Capt'n John: Story of a Texas Ranger*, Nortex Press, San Antonio 1991.

Gold, Ella, Translator, *Kirchen—Buch: Church Record Book of the Vereinskirche 1849-1870*, Published by Gillespie County Historical Society, Inc., Fredericksburg, Texas, 1986.

Glenn, Frankie Davis, *Frontier Series: John William Sansom's Battle of the Nueces*, Published by Frankie Davis Glenn, Boerne, Texas, 1991

Gurasich, Marj, *A House Divided*, Texas Christian University Press, Fort Worth, Texas, 1994.

Harper Centennial Committee, *Here's Harper*, Radio Post, Inc., Fredericksburg, Texas, 1963.

Harper Texas Sesquicentennial Committee, *Here's Harper Two*, Nortex Press, Austin, Texas, 1986.

Jordan, Terry C., *German Seed in Texas Soil; Immigrant Farmer in Nineteenth-Century Texas*, University of Texas Press, Austin, Texas 1966.

Lonn, Ella, *Foreigners In The Confederacy*, The University of North Carolina Press, Chapel Hill, North Carolina, 1940.

Lossing, Benson J, LL.D, P*ictorial Field Book of The Civil War*, 1997, Volume II, 1997.

McGowen, Stanley S., *Horse Sweat and Power Smoke: The First Texas Cavalry in the Civil War*, Texas A & M Press, College Station, Texas 1999

Menard County Historical Society, Ed, *The Menard County History: An Anthology*, Anchor Publishing Company, San Angelo, Texas, 1982.

Michener, James, *Texas*, Random House, New York, New York, 1985.

Nunn, W. C., *Escape From Reconstruction*, Texas Christian University, Fort Worth, Texas, 1956.

Olmstead, Frederick Law, *A Journey Through Texas Or, A Saddle-trip on the Southwestern Frontier*, Austin, Texas: University of Texas Press, 1978.

Pirtle, Caleb III and Cusack, Michael F., *Fort Clark: The Lonely Sentinel On Texas's Western Frontier*, Eakin Press, Austin, Texas, 1985.

Ruggero, Ed, *Combat Jump: The Young Men Who Led the Assault into Fortress Europe*, Harper Collins, New York, New York, 2003.

Simpson, Colonel Harold B., & Wright, Brigadier General Marcus J., *Texas In The War 1861—1865*, The Hill Junior College Press, 1965.

Smith, David Paul, *Frontier Defense in the Civil War*, College Station, Texas: Texas A & M University Press, 1992.

Smith, Thomas C., *Here's Yer Mule-The Diary of Thomas C. Smith, 3rd Sergeant, Company G. Wood's Regiment, 32nd Texas Cavalry* (AKA 36th Texas Cavalry), The Little Texas Press, Waco, Texas, 1958.

Smith, Thomas Tyree, *Fort Inge, Sharps, Spurs, and Sabers on the Texas Frontier 1849—1869*, Eakins Press, Austin, Texas.

Stanley, D. S., *Personal Memories of Major General D. S. Stanley*, Harvard Press, Cambridge, Massachusetts, c1917.

Stewart, Mike and Anne, *Comfort Women in Comfort History*, Privately published by Anne and Mike Stewart, Comfort, Texas, 1993.

Steward, Mike and Anne, *Death on the Nueces: Mina Stieler Stories*, Privately published by Anne and Mike Steward, Radium Springs, New Mexico 1997.

Tolzmann, Don Heinrich, Ed., *The German-American Forty-Eighters 1848-1998*, Indiana University Press, Indianapolis, Indiana, 1998.

Turner, Mary Lewis, *Julius Theodore Splittgerber 1819—1897, Volume Two: His German Ancestors and American Descendants*, Watercress Press, San Antonio, Texas, 1997.

Weber, Adolf Paul Weber, *Deutsche Pioniere, Zur Geschichte Des Deutschthums in Texas*, Selbstverlag Des Verfasserrs, Published by Paul Weber, San Antonio, Texas 1894.

Webster's New Collegiate Dictionary, A Merriam-Webster G. & C. Merriam Company, Springfield, Massachusetts, 1979.

Wooster, A., *Texas and Texans in the Civil War*, Eakin Press, Austin, Texas, 1995.

Thompson, Leroy, *The Counter Insurgency Manual*, The Military Book Club, Stackpole Books, Mechanicsburg, Pennsylvania, 2002.

Tyler, Ron, et al, Ed, *The New Handbook of Texas*, Six Volumes, Austin, Texas: Texas State Historical Association, 1996.

Webb, Walter P., Ed., *The Handbook of Texas, Volume I & II*, Austin, Texas: Texas State Historical Association, 1952.

Weber, Adolf Paul, *Die Deutsche Pioniers Zur Geschichtes des Deutschthums in Texas*.

ORAL ACCOUNTS, INTERVIEWS AND PRESENTATIONS

Schmidt, Eduard, English Translation of Address Commemorating The 50th Anniversary Of The Battle On The Nueces, August 1862, copy provided by Gregory J. Krauter, Comfort, Texas.

Mary Lewis Turner's oral presentation to the German-Texas Heritage Society at its September 5-7, 1997 Convention at

Kerrville, Texas and off the record interview, July 19, 1997 at Kerrville, Texas.

Oral interview with Amelia "Mollie" Eckhardt Dennis, an August Duecker descendent, September 5, 1997.

JOURNALS

Barr, Alwyn (Ed), "Records of the Confederate Military Commission in San Antonio July 2 — October 10, 1862" (CMC), Alwyn Barr, 'Southwestern Historical Quarterly', Volumes LXX (July 1966); LXX (October 1966); LXX (April 1967); LXXI (October 1967); LXXIII (July 1969).

Baulch, Joe, "The Dogs of War Unleashed: The Devil Concealed in Men Unchained" *West Texas Historical Association Year Book*, Volume LXXIII, 1997.

Biesele, Rudolph L., "The Texas State Convention of Germans in 1854", *Southwestern Historical Quarterly*, Volume XXXIII, April 1930.

Boerner, Dr. Jur Bernhard (An English Translation) "The Bonnet Family From Chambons In The Dauphine." Deutfsches Geschlechterburh, Volume 60, 1, 1928.

Clare, Mary, "Bloody Ground: The Incident on the Nueces", *Civil War*, Issue Number 70, October 1998.

Elliott, Claude, "Union Sentiment in Texas 1861- 1865", Southwestern Historical Quarterly, Volume L, January 1947.

"German Unionists in Texas", *Harper's Weekly*, New York, New York, January 20, 1866.

Gold, Gerald R., "Gillespie County in the Civil War", *The Junior Historian*, May 1965.

Hall, Ada Maria, *The Texas Germans in State and National Politics, 1850 – 1865*, Austin, Texas: M. A. Thesis, University of Texas, 1938.

Kamphoefner, Walter D. 'New Perspectives on Texas Germans and the Confederacy by Walter D. Kamphoefner, *Southwestern Historical Quarterly*, Volume CII No. 4 (April 1999).

Kelton, Elmer, "The Fleeing Sixty-A True Story", *Ranch Romances Magazine*, January 18, 1952.

Lossing, Benson J, LL.D, *Pictorial Field Book of The Civil War*, 1997, Volume II, 1997.

Nixon, Victor, "An Encounter With The Partisan Rangers", *The Junior Historian*, Texas State Historical Association, Austin, Texas, May 1, 1965.

Rutherford, Phillip, "Defying The State of Texas", *Civil War Times Illustrated*, Volume 19, No. 1, April 1979.

Sansom, John W., "The German Citizens Were Loyal To The Union", *Hunter's Magazine*, II, November 1911.

Schutze, Albert, "Was a Survivor of the Nueces Battle", *Frontier Times*, Volume Six, Number Eight, October 1924.

Shook, Robert W., "The Battle of the Nueces, August 10, 1862", *Southwestern Historical Quarterly*, Volume LXV, October 1961.

Thorpe, Helen, 'Historical Friction", *Texas Monthly*, Austin, Texas October 1997.

UNPUBLISHED LETTERS, MANUSCRIPTS, MISCELLANEOUS, JOURNALS & DIARIES

Betzer, Roy J. *Early Fredericksburg and Fort Martin Scott*, an unpublished manuscript.

Comfort Heritage Foundation, Inc., Handout, n. d.

Souvenir Book of Comfort, Texas Commemorating 75th Anniversary August 13, 1929 of The Battle of Nueces.

Journal, Fritz Schellhase, c1912.

Ledger of John Sansom, Sansom file, Daughter of Republic of Texas Library at the Alamo, San Antonio, Texas.

Letter, August Hoffmann to his children, September 1, 1925.

Letter, Captain Henry Davis, July 25, 1862, to Colonel James Norris, A.G.C., TSA, Austin, Texas.

Letter, Captain Charles de Montel, Camp Verde, August 3, 1862, to Colonel James Norris, A.G.C., Austin, Texas.

Letter, Catherine Carrigan to Gregory Krauter, Comfort, Texas, August 1, 1991.

Letter, Ernst Cramer to 'My Beloved parents', Monterrey, Mexico, October 30, 1862.

Letter, D. H Farr, Kerrsville, Kerr County, February 13, 1862 to Governor Lubbock, Governor Lubbock's File, Texas State Archives (TSA), Austin, Texas.

Letter, Eduard Degener, August 1, 1861 to his father-in-law Colonel Von Bernewitz.

Letter, Fritz Tegener, Austin, Texas, to Herr August Duecker, Gillespie County, Texas, August 23, 1875.

Letter, Henry Schwethelm, October 1913, to his grandson Otto.

Letter, Howard Henderson, October 1908, Ingram, Texas, contained in Ransleben, Guido E., *A Hundred Years of Comfort in Texas*, The Naylor Company, San Antonio, Texas, 1974

Letter, J. W. Seal, September 8, 1862.

Letter, Jacob Kuechler to James Newcomb, August 1887, read at the Twenty-Fifth Anniversary of Nueces Battle.

Letter, Jacob Kuechler, as quoted by Ransleben in *A Hundred Years of Comfort in Texas*, The Naylor Company, San Antonio, Texas 1954.

Letter, Jacob Kuechler, Toyah, County, October 26, 1888 to "Dearest Marie"

Letter, John Donelson, Provost Marshal, Camp Pedernales, August 21, 1862.

Letter, John W. Sansom to James T. De Shields, August 14, 1907.

Letter, Julius Schlickum to his father-in-law, dated December 21, 1862, on board the English frigate *Hope*.

Letter, Frank V. D. Stucken, Fredericksburg, February 13, 1862, to Governor Lubbock, Governor Lubbock's File, TSA, Austin, Texas.

Letter, R. G. Gibson, Camp Davis, March 31, 1864, A.G.C., TSA, Austin, Texas.

Letter, T. R. Fehrenbach, San Antonio, Texas March 27, 1997 to author.

Petition, Citizens of Kerr County, To Governor Lubbock, dated February 14, 1862, Governor Lubbock's File, TSA, Austin, Texas.

Usener, Raymond, The Jacob and Ludwig Usener Story, March 1997.

Schmidt, Eduard, English Translation of Address Commemorating The 50th Anniversary Of The Battle On The Nueces, August 1862.

Schweppe, F. W., "Bonnet Brothers", copy located in the Edith A. Gray Library, Boerne Public Library, Boerne, Texas, copy provided by Ester Bonnet Strange, Kerrville, Texas.

Siemering, August, *Ein Verstehl Tex Leben*, Published by August Siemering, San Antonio, Texas 1876

Siemering, August, *Des Lateimische in Texas*, Published by August Siemering, San Antonio, Texas, 1874.

Statement, John Larremore, April 2, 1864, Gillespie County Court Records.

Striegler, Selm, *The Striegler Family History*, n.d.

Tombstone, Heinrich and Jakob Itz, Der Friedhof Cemetery, Fredericksburg, Texas, Section 18, Graves 12 and 13.

PAPERS, THESES, AND DISSERTATIONS.

Curtis, Sara Kay, *A History of Gillespie County, Texas 1846—1900*, Austin, Texas, M. A. Thesis, University of Texas, Austin, Texas 1943.

Dykes-Hoffmann, Judith, *Treue Der Union: German Texan Women On The Civil War Homefront*, San Marcos, Texas: M. S. Thesis, Southwest Texas State University, 1996.

Felger, Robert Pattison, *Texas In The War For Southern Independence 1861 – 1865*, Austin, Texas: Ph. D. Dissertation, University of Texas, 1947.

Hall, Ada Maria, *The Texas Germans in State and National Politics, 1850-1865*, Austin, Texas: M. A. Thesis, University of Texas, 1938.

Heintzen, Frank W., *Fredericksburg, Texas During The Civil War And Reconstruction*, San Antonio, Texas: M. A. Thesis, St. Mary's University, 1944.

Henderson, Shawn, *A Descriptive Case Study of the Encounter at the Nueces River on August 10, 1862*, M. A. Thesis, Urbana University, Urbana, Ohio 2007.

Johnson, Melvin C., *A New Perspective for the Antebellum and Civil War Texas German Community*, M. A. Thesis, Stephen F. Austin State University, Nacogdoches, Texas 1993.

Weinheimer, Ophelia Nielsen, *The Early History of Gillespie County*, San Marcos, Texas: M. A. Thesis, Southwest Texas State Teachers College, 1952.

NEWSPAPERS

Alberthal, Vernel, "Bushwhackers In Them Thar Hills", *The Radio Post*, October 2, 1952.

Austin Semi-Weekly News, August 29, 1861.

Austin State Gazette, June 8, 1861.

Austin State Gazette, August 31, 1861.

Biffle, Kent, 'Remembering Hill County Bad Old Days" *Dallas Morning News*, November 23, 1997.

Neu Braunfelser Zeitung, September 6, 1862.

Raley, Helen "Blackest Crime in Texas Warfare", *Dallas Morning News*, May 5, 1929.

Report, Lieutenant Colonel Nat Benton, *San Antonio Weekly Herald*, August 30, 1862.

San Antonio Daily Ledger and Texan, August 29, 1861.

San Antonio Herald, May 31, 1862.

San Antonio Herald, June 8, 1861.

San Antonio Herald, June 11, 1862.

San Antonio Weekly Ledger and Texan, August 31, 1862.

Siemering, August, "Die Deutschen in Texas Waehrend Des Buergerkrieges (The Germans in Texas during the Civil War)", *Freir Presse fuer Texas*, May and June 1923.

"The Diary of D. P. Hopkins", *San Antonio Express*, January 13, 1918.

www.ingramcontent.com/pod-product-compliance
Lightning Source LLC
Chambersburg PA
CBHW030514100426

42813CB00001B/40